D1203076

THE GEOMETRIC TOPOLOGY
OF 3-MANIFOLDS

AMERICAN MATHEMATICAL SOCIETY
COLLOQUIUM PUBLICATIONS

VOLUME 40

THE GEOMETRIC TOPOLOGY OF 3-MANIFOLDS

BY

R. H. BING

PROFESSOR OF MATHEMATICS
UNIVERSITY OF TEXAS AT AUSTIN

AMERICAN MATHEMATICAL SOCIETY
PROVIDENCE, RHODE ISLAND

1980 *Mathematics Subject Classification.* Primary 57–XX.

Library of Congress Cataloging in Publication Data

Bing, R. H., 1914–
    The geometric topology of 3-manifolds.
    (Colloquium publications; ISSN 0065-9258; v. 40)
    Bibliography: p.
    Includes index.

    1. Geometric topology. 2. 3-manifolds.
I. Title. II. Series: Colloquium publications (American Mathematical Society); v.40.
QA612.S94   1983   514.2   83-14962
ISBN 0-8218-1040-5

TABLE OF CONTENTS

PREFACE

This book contains an elaboration of some aspects of my Colloquium Lectures given before the American Mathematical Society in 1970. It represents a continuation of my efforts to bridge the gap between the study of topological objects in 3-space and the corresponding polyhedral objects.

One of the main goals is to give an understandable proof of the side approximation theorem. This theorem was introduced in the early 60's and has been useful in studying topological surfaces. There have been alternative proofs of it suggested in research papers but perhaps none so extensive as that included in Chapters X–XIII. I have considered several versions of this proof but doubt that the best treatment is yet at hand.

Topics related to the side approximation theorem include wild surfaces (Chapters IV, VI), the Schoenflies theorem (Chapters III, V, XIV), Dehn's lemma (Chapters XV, XVI), and the loop theorem (Chapter XVII). Applications of these topics are listed in Chapter XVIII.

So as to make the book somewhat self-contained, some preliminary results from PL topology, homotopy theory, and homology theory that are of particular use in the study of 3-manifold are treated in Chapters I, II, VII, VIII, IX. The material is traditional but it is treated in the down-to-earth context in which it is applied. This provides a straightforward approach for those who like to develop mathematics from the specific to the general.

I was very fortunate to have Professor R. L. Moore as a teacher. His success in leading students to formulate theorems and proofs is legendary. During these student days I profited from knowing such mathematical leaders as E. E. Moise, F. B. Jones, R. D. Anderson, C. E. Burgess, Mary Ellen Rudin, Billy Jo Ball, Mary Elizabeth Hamstrom. Later, as colleagues, Ed Floyd, Ed Fadell, Deane Montgomery, Joe Martin, Bill Eaton, Steve Armentrout, Jim Cannon, and Mike Starbird had a big impact on my research activities. Work with graduate students has been very stimulating, and in particular, that with Mort Brown, Russ McMillan, Don Sanderson, Bob Daverman, David Gillman, David Henderson, and John Hempel is relevant to the contents of this book. Graduate students Robert L. Dawes, Russell Rose, and Richard Skora were helpful in reviewing versions of the book.

My mother had mathematics as a hobby, and I learned from her even before starting school that arithmetic was fun if done rapidly and accurately. She later encouraged the notion that geometric proofs were to be discovered and proved rather than learned.

In addition to all this mathematical help, I received financial research support from the University of Wisconsin, The University of Texas, and the National Science Foundation. I also recognize the encouragement of my wife, Mary, which made it possible for me to spend so much time with mathematics and students.

R. H. BING

CHAPTER I

PLANAR COMPLEXES

Some may feel that a 3-manifold locally looks like a stack of planes and that is about all the local structure it has. However, if one contemplates the matter further, one sees that a 3-manifold contains many strange objects and there are questions as to their properties. How do they intersect each other? How can they be reimbedded? What do they bound? Gaze into R^3 and contemplate its topological and geometric mysteries.

Three chapters on the plane provide a springboard for a study of the local geometric topological properties of 3-manifolds. A knowledge of the topology of the plane R^2 is useful in the study of 3-manifolds. Many theorems about 3-manifolds are merely extensions of planar theorems that have similar proofs. We prefer the approach of considering such theorems first in a low-dimensional setting before proving the related results in higher dimension. In case the higher-dimensional approach differs little from the 3-dimensional result, we may even skip a detailed treatment of it.

If J is a simple closed curve in the plane R^2, then J separates R^2 as shown in Chapter III. This theorem generalizes to higher dimensions (Chapter VII). Although there is a homeomorphism of R^2 onto itself taking J to a circle (Chapter III), this result does not extend to higher dimensions (Chapter IV). These are examples of topological results we shall consider. Straight lines, rectilinear simplexes (considered later in Chapter I), PL maps (Chapter II), and the PL Schoenflies theorem (Chapter III and XIV) are tools. It is our aim to ultimately get topological results, but we prove theorems about our tools first—not as a goal in itself, but to help get these topological results later. The theorems of this first chapter are tools that can be used to give topological results. They generalize from the plane to higher dimensions.

While we prove some basic results about the plane, we do not start at the beginning nor give a complete axiomatic treatment of geometry. Rather, we assume as true things known from elementary courses, analytics, calculus, geometry, or observations. A line separates the plane into two pieces; a triangle

1

(sometimes called a triangular polygon to distinguish it from a 3-sided disk) separates the plane into two pieces; if in Euclidean space p is a point and X is a closed set, there is a point of X as close to p as any other point of X. In later chapters we use more sophisticated results as the needs arise.

Thinking mathematicians find proofs of theorems easier and more meaningful if they think of the proofs themselves rather than read them. Although the proofs of some theorems are included, it is recommended that the user of this book first think about the results, use diagrams to help understand the situations, and try to find the proofs themselves. They might read the proofs given as a last resort. We follow the convention of putting a $\square$ at the end of a proof—or even at the end of a stated theorem if no proof is to be included.

I.1. Triangulations. A triangular disk is called a 2-*simplex*. The faces of a 2-simplex Δ^2 are the edges of Δ^2, the vertices of Δ^2, and Δ^2 itself. Hence Δ^2 has seven faces. It has eight faces if one uses the convention of the empty set and regards this "empty set" as a face.

In general, an *n-simplex* $\Delta^n = v_0 v_1 \cdots v_n$ *in Euclidean n-space* R^n is the convex hull of $\cup \{v_i\}$ where $v_0, v_1, \ldots, v_n$ are $n + 1$ points in R^n such that no $(n - 1)$-hyperplane in R^n contains all of these points. We say that Δ^n is of dimension n. The convex hull of each subcollection of $\cup \{v_i\}$ is a *face* of Δ^n. Any set isometric to an n-simplex in R^n is called an *n-simplex*. The isometry determines its *faces*. (Recall that an isometry is a distance preserving homeomorphism.) Note that the faces of a simplex are simplexes. How many faces does an n-simplex have?

An object homeomorphic to an n-simplex is sometimes called an *n-ball* or *n-cell*. We frequently use D to denote a 2-cell (thinking D for disk) and B to denote a 3-cell (thinking B for ball).

The union of the $(n - 1)$-dimensional faces of a simplex Δ^n is called the boundary of Δ^n and denoted by Bd Δ^n. The interior of Δ^n (denoted by Int Δ^n) is $\Delta^n -$ Bd Δ^n. For a 1-simplex $v_i v_j$, we sometimes denote Int $v_i v_j$ by $(v_i v_j)$.

Suppose R^2 is subdivided into a locally finite collection of triangles such that if two of these triangles intersect, the intersection is an edge or vertex of each. This subdivision determines a *triangulation T* of R^2. The elements of T are the triangles (2-simplexes), edges of the triangles (1-simplexes); and vertices (0-simplexes). If T_1, T_2 are triangulations of R^2, we say that T_2 is a *subdivision* (or *refinement*) of T_1 if each simplex of T_2 is a subset of an element of T_1. We say that T_2 was obtained by *subdividing T_1*.

Not all metric spaces are the unions of locally finite collections of simplexes, but some are. If X is such a locally finite union we call T a *triangulation* of X if it is a collection of simplexes such that

X is the union of the elements of T,

each face of an element of T is an element of T,

if two elements of T intersect, the intersection is a face of each, and

T is locally finite at each point of X.

We call (X, T) a *complex* where T is a triangulation of the metric object X. The same object may have many triangulations. We say that (X_1, T_1) is a subcomplex of (X, T) if each simplex of T_1 is an element of T. We call (X, T) an *i-complex* if T's simplex of highest dimension is an *i-simplex*.

At times we need to consider a triangulation of an object (such as a round ball) not made up of ordinary simplexes. If there is a homeomorphism h of an object X onto a complex C with a triangulation $T(C)$, we may regard X as having a *curvilinear triangulation* whose simplexes are the inverses under h^{-1} of the simplexes of $T(C)$. However, if we want it clearly understood that we are not considering curvilinear simplexes but only ordinary linear ones, we may emphasize this by saying that the triangulation is *rectilinear*. The use of the word "rectilinear" is usually redundant here since it is understood to hold without mentioning it unless the context suggests otherwise.

The *i-skeleton* of a triangulation T is the union of the *i*-simplexes of T. We denote the *i*-skeleton by T^i. It may be noted that we follow the geometric rather than the algebraic point of view and regard skeletons and boundaries as subsets of metric spaces rather than as collections of faces.

THEOREM I.1.A. *Any two triangulations of the plane have a common refinement. The refinement can be chosen so that its vertices lie in the union of the 1-skeletons of the two given triangulations.*

A clue as to how to start building the common refinement follows from considering a 2-simplex Δ^2 of the first triangulation and the 1-skeleton T^1 of the second. Construct a triangulation of Δ^2 so that the 1-skeleton of this triangulation contains $\Delta^2 \cap T^1$. So that this triangulation of Δ^2 agrees with a similar triangulation of an adjacent 2-simplex, it might be prudent to pick the triangulation of Δ^2 to have a minimum number of vertices on Bd Δ^2. □

Theorem I.1.A can be extended to get the following result.

THEOREM I.1.B. *Any two rectilinear triangulations of the same geometric object in R^2 have a common refinement there.* □

THEOREM I.1.C. *If T is a triangulation of a closed subset X of R^2, there is a triangulation T_1 of R^2 such that the 1-skeleton of T is a subset of the 1-skeleton of T_1.*

The proof of Theorem I.1.C follows from the techniques of the proof of Theorem I.1.B. □

THEOREM I.1.D. *Any triangulation T of a closed subset C of R^2 can be extended to a triangulation T_1 of R^2 such that $T \subset T_1$.*

PROOF. Note that we are not permitted to subdivide the elements of T. To prevent a possible subdivision we shield certain exposed 1-simplexes of T. A 1-simplex Δ^1 of T is called *exposed* if each point of Δ^1 is a limit point of $R^2 - C$. If Δ^1 is exposed from only one side there is a 2-simplex Δ^2 such that $\Delta^2 \cap C = \Delta^1$ and Δ^1 is not exposed in $\Delta^2 \cup C$. We say that Δ^2 shields Δ^1. If Δ^1 is exposed from both sides, two 2-simplexes can be used to shield Δ^1. Suppose enough such 2-simplexes are added to C to shield all exposed 1-simplexes in C. Denote the union of C and these shielding 2-simplexes by C_1 and the resulting triangulation by T_2. It is assumed that no two of the shielding 2-simplexes intersect each other except possibly at a shielded face or a vertex of such a face.

It follows from Theorem I.1.C that there is a triangulation T_3 of R^2 such that each simplex in T_2 is the union of simplexes in T_3. The simplexes of T are elements of T_1. It would be simplistic to let the elements of T_3 in the closure of $R^2 - C$ be elements of T_1 for there might be a mismatch on 1-simplexes of T that were originally exposed. Instead we assign elements of T_3 in the closure of $R^2 - C_1$ to T_1 and note that we have now triangulated all but the shielding 2-simplexes. For each such shielding Δ^2, we have already triangulated Bd Δ^2 and we complete the triangulation T_1 by coning Bd Δ^2 from a point of Int Δ^2. $\quad\square$

I.2. Extending triangulation. One of the easy things about the study of 3-manifolds is that many of the proofs of 3-dimensional results are minor variations of 2-dimensional results. The fact that not all such 3-dimensional claims are such minor extensions (or even true) adds spice. Some might prefer to prove a theorem in complete generality and then note that the general theorem implies its truth in each special case. However, it frequently adds to the understanding of a result to see first how it is proved in special easy cases.

Theorem I.1.D is a planar extension theorem for R^2. We show how it can be extended to give an extension theorem about R^3. If one has a rectilinear triangulation of a closed subset of R^3 (or R^n) it is sometimes useful to extend this triangulation to a rectilinear triangulation of R^3 (or R^n).

THEOREM I.2.A. *If T is a triangulation of a closed subset C of R^3, then there is a triangulation T_1 of R^3 such that each simplex of T is a simplex of T_1.*

PROOF. The proof is a variation of that of Theorem I.1.D and is given in three steps.

Step 1. Shielding the 2-simplexes of T. Suppose Δ^2 is a 2-simplex of T. If no point of Int Δ^2 is a limit point of $R^3 - C$, Δ^2 is already shielded. If Int Δ^2 is accessible from only one side in $R^3 - C$, we cover the interior of this side by adding to C a 3-simplex Δ^3 having Δ^2 as a face such that $\Delta^3 \cap C = \Delta^2$. If Int Δ^2 is accessible from both sides in $R^3 - C$, we add two 3-simplexes to C to cover Int Δ^2. We shield each 2-simplex of T in this fashion and obtain a closed set C_1 with a rectilinear triangulation T_2 such that each simplex of T is a simplex of T_2 and each 2-simplex Δ^2 in T is shielded in C_1. It is supposed that the shielding 3-simplexes do not intersect each other unnecessarily.

Step 2. Shielding 1-simplexes of T. After the 2-simplexes are shielded we turn our attention to the 1-simplexes. Suppose Δ^1 is a 1-simplex of T. If no point of Int Δ^1 is accessible from $R^3 - C_1$, Δ^1 is already shielded. If Δ^1 is not an edge of any 3-simplex of T_2, we augment C_1 by adding three 3-simplexes with Δ^1 as a common edge. When this is done for each bare 1-simplex we call T_3 the triangulation of the resulting C_2. If Int Δ^l is accessible from $R^3 - C_2$, it is accessible through certain dihedral angles whose sides contain 2-simplexes of T_3 with Δ^l as a common edge. Figure I.2.A shows how the 3-simplexes of T_3 having Δ^l as an edge might fit. This model shows one large dihedral angle (on the back side) and one small (on the front), but that is not important.

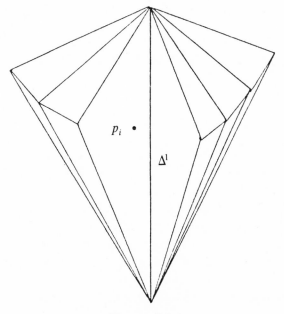

FIGURE I.2.A

Put a point p_i near the middle of Δ^l on the bisector of each vacant angle and cone from p_i over the two adjacent 2-simplexes of T_3. Adding two such 3-simplexes for each vacant angle blocks the accessibility of Int Δ^l through the dihedral angles. When all angles are blocked, we use T_4 to denote the triangulation of the resulting closed set C_3.

Step 3. *The triangulation* T_1. We shall define T_1 on C, then on the closure of $R^3 - C_3$, and finally on the buffer zone. First, T_1 agrees with T on C.

A theorem whose proof resembles those of Theorems I.1.B and I.1.C shows that there is a triangulation T_5 of R_3 such that each simplex of T_4 is the union of simplexes of T_5. Then T_1 agrees with T_5 on the closure of $R^3 - C_3$.

There is no 1-simplex Δ^l of T_4 whose interior is in the buffer. If Δ^2 is a 2-simplex of T_4 whose interior is in the buffer, we have already defined T_1 on Bd Δ^2 and could assign T_1 on Δ^2 by coning Bd Δ^2 from a point of Int Δ^2. Also, if Δ^3 is a 3-simplex of T_4 with interior in the buffer, T_1 can be determined on Δ^3 by coning. $\square$

The preceding theorem generalizes to many situations. In fact, there may not actually be an optimal generalization in the sense that no matter what generalization is given, someone could produce a more general one. One of the advantages of being an understander rather than a theorem quoter is that one may be able to obtain approaches to a wide variety of theorems some of which may not even have been formulated yet.

The techniques used to prove Theorems I.1.D and I.2.A can be used to prove the following.

THEOREM I.2.B. *If (X, T) is a complex and U is an open subset of X, then there is a triangulation T_U of U such that each simplex of T_U lies in a simplex of T.*   □

I.3. Special triangulations. If Δ^n is an *n*-simplex and $p \in \text{Int } \Delta^n$, we can obtain a triangulation T of Δ^n by *coning* Bd Δ^n from *p*. Faces of Δ^n of dimension less than *n* are elements of the triangulation *T*. The point *p* is a vertex of *T*. If Δ^j is a face of Δ^n with $j < n$ the join *p* and Δ^j is an element of *T*. The *join* of two objects *A*, *B* is the union of all 1-simplexes with one end in *A* and the other in *B* if *A* and *B* are arranged so that no two of the 1-simplexes intersect except possibly at an end point of each. If *p* is a point , the join of *p* and *X* is called the cone over *X* from *p*.

A subdivision of a complex (X, T) that has attracted special attention is a *stellar subdivision*. Under it only one new vertex is used. If $p \in X$ and *p* is not a vertex of *T*, it lies on the interior of some face—say $p \in \text{Int } \Delta^j$. We obtain the stellar subdivision T' by coning Bd Δ^j from *p*, leaving the other *i*-simplexes $(i \leq j)$ alone and then proceeding to subdivide the simplexes of *T* with dimensions more than *j*. If Δ^{j+1} is a $(j + 1)$-simplex of *T* whose boundary was not subdivided ($\Delta^j \not\subset \text{Bd } \Delta^{j+1}$), $\Delta^{j+1} \in T'$. If Bd Δ^{j+1} contained a subdivided face $(\Delta^j \subset \text{Bd } \Delta^{j+i})$, it contains a vertex *q* not on the face and the subdivision of Δ^{j+1} is obtained by coning the subdivided face from *q*. Continuing in this fashion the $j + 2, j + 3,\ldots,n$ simplexes of *T* are subdivided or inserted into T' intact.

At one time there was considerable hope of being able to handle most subdivision problems through the stellar route. However, a snag developed when researchers were unable to show that any two triangulations of the same rectilinear compact object had a common stellar subdivision. See Theorem I.1.B.

EXERCISE I.3.A. Suppose *T* is a triangulation of R^2, Δ^1 is a 1-simplex of *T*, $p \in \text{Int } \Delta^1$, and T' is stellar subdivision of *T* obtained by making *p* a vertex. How do the number of 0-simplexes, 1-simplexes, and 2-simplexes of T' compare with the number in *T*?

Answer. One more, three more, two more.   □

If Δ^j is a *j*-simplex with vertices $v_0, v_1,\ldots,v_j$, the barycenter of Δ^j is the center of mass of equal weights assigned to the vertices of Δ^j. As pointed out in §1 of the next chapter, we could represent the point as $(v_0 + v_2 + \cdots +v_j)/(j + 1)$. If (X, T) is a complex, the *barycentric subdivision* T' of *T* is obtained as follows: first, subdivide each 1-simplex of *T* by chopping that 1-simplex at the barycenter of the 1-simplex; next, subdivide the 2-simplexes by coning from the barycenter of the 2-simplex over the already triangulated boundary of the 2-simplex which was considered at the preceding step; next, subdivide each 3-simplex by coning from the barycenter over the already triangulated boundary; etc. If T' is the barycentric subdivision of *T* and T'' is the barycentric subdivision of T', we call

T'' the *second barycentric subdivision* of T. It is sometimes just called the *second derived*.

EXERCISE I.3.B. If T'' is a second barycentric subdivision of some finite 2-complex (X, T) how does the number of 2-simplexes in T'' compare with the number in T?

Answer. 36 times as many. □

Chapter I contained only a brief treatment of some basic properties of triangulations. An understanding of how to work with such tools and prove theorems about them is valuable. They offer material on which to cut one's eye-teeth before passing quickly to more interesting results.

Chapter II

PL PLANAR MAPS

There is merit in moving rapidly through the first two chapters. The theorems are for background and are useful tools. Our familiarity with the plane causes us to introduce many theorems in that space. In most cases the theorems extend to higher dimensions by the same techniques, and we will not repeat the proofs.

II.1. Linear maps. A continuous function or transformation is called a *map*. A map of simplex Δ into Euclidean space (or another simplex) is called *linear* if f preserves linearity and ratios. We say that f *preserves linearity* if straight-line segments in Δ are sent by f into straight-line segments (or points). It *preserves ratios* if for each three collinear points x_1, x_2, x_3 of Δ, $f(x_2)$ divides $f(x_1), f(x_3)$ in the same ratio that x_2 divides x_1, x_3—that is, $d(x_1, x_2) \cdot d(f(x_2), f(x_3)) = d(x_2, x_3) \cdot d(f(x_1), f(x_2))$ where d denotes Euclidean distance. Projection maps preserve linearity, but some do not preserve ratios and are hence not linear. Figure II.1 shows the projection of triangle abc into a plane π from a point p. It is not true that a point halfway between a and b projects into a point halfway between a' and b'.

The above definition of a linear map is geometric. There is also the physical approach where a map f of a simplex into Euclidean m-space R^n is defined to be linear if the center of gravity of weights hung onto the vertices of the simplex is sent by f to the center of gravity of the same weights hung on the corresponding images of the vertices. We shall not use this physical approach but mention it for its visual appeal.

There is also the algebraic approach where we recall from algebra that $a_0 + a_1 x_1 + \cdots + a_n x_n$ is a linear function on n variables (the a's are constant and the x's variables). It is convenient to denote a point x of R^n by its coordinates $(x_1, x_2, \ldots, x_n)$ so we write $x = (x_1, x_2, \ldots, x_n)$. A map f of a subset X of R^n into R^m is defined to be linear if there are m linear functions $f_i(x) = a_{i0} + a_{i1} x_1 + \cdots + a_{ij} x_n$ $(i = 1, 2, \ldots, m)$ such that for each $x \in X, f(x)$ has coordinates $f_1(x), f_2(x), \ldots, f_m(x)$.

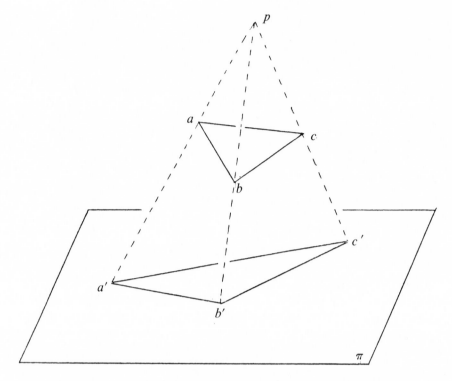

Figure II.1

If p_1, p_2 are planar points with coordinates (x_1, y_1), (x_2, y_2) and a_1, a_2 are constants, we use $a_1 p_1 + a_2 p_2$ to denote the point with coordinates $(a_1 x_1 + a_2 x_2, a_1 y_1 + a_2 y_2)$. Each point $p \in v_1 v_2 v_3$ can be written as $a_1 v_1 + a_2 v_2 + a_3 v_3$ where a_1, a_2, a_3 are nonnegative numbers whose sum is 1. Exercise II.1.A shows one way to find a_1, a_2, a_3 if v_1, v_2, v_3, p are given.

EXERCISE II.1.A. Suppose f is a map of triangular disk $v_1 v_2 v_3$ into triangular disk $v_1' v_2' v_3'$ described as follows: $f(v_i) = v_i'$ ($i = 1, 2, 3$); if x is between v_2 and v_3, $f(x)$ is the point between v_2' and v_3' that divides v_2', v_3' is the same ratio that x divides v_2, v_3; if x is between v_2 and v_3 and y is between v_1 and x, then $f(y)$ is the point between v_1' and $f(x)$ that divides v_1', $f(x)$ in the same ratio that y divides v_1, x. Show that if the ratios are $t: 1 - t$ and $s: 1 - s$ respectively as suggested by Figure II.1.A and the points are planar vectors, then

$$x = tv_3 + (1 - t)v_2,$$
$$y = (1 - s)v_1 + stv_3 + s(1 - t)v_2, \quad \text{and}$$
$$f(y) = (1 - s)v_1' + stv_3' + s(1 - t)v_2'. \quad \square$$

While one could use equations to show that various definitions of linearity are equivalent, it is to be admitted that geometric topologists are more concerned with concepts than with equations. A frequently used basic concept is that under a linear map the image of a simplex is the convex union of a finite number of simplexes.

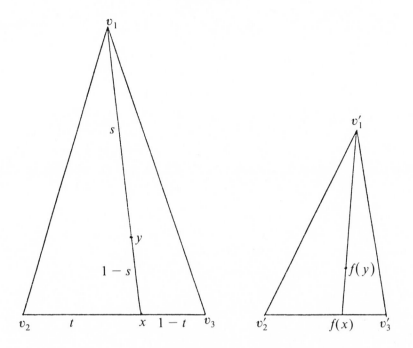

Figure II.1.A

II.2. PL maps. If (X, T) is a complex, we call a map f of X into R^2 (or R^n) a *linear* map of (X, T) if f is linear on each simplex of T. A map of a complex (X, T) into complex (Y, S) is *linear* if f takes each simplex of T linearly *into* a simplex of S. (Some authors insist that f takes each simplex of T linearly *onto* a simplex of S but we do not.) A map f of a complex (X, T) into a complex (Y, S) is *piecewise linear* (PL) if there is a subdivision t_1 of T such that f takes each simplex of T_1 linearly into a simplex of S.

THEOREM II.2.A. *Suppose ϕ is a linear map of a 2-simplex Δ into R^2 and T_2 is a triangulation of $\phi(\Delta)$ such that the image of each face of Δ is the union of elements of T_2. Then there is a triangulation T of Δ such that ϕ takes each simplex of T linearly onto a simplex of T_2.*

PROOF. If $\phi(\Delta)$ is a point, any subdivision of Δ would serve for T, but to be specific, we pick the one where $\Delta \in T$.

If ϕ is a homeomorphism, we describe T by requiring that for each $\sigma \in T_2$, $f^{-1}(\sigma) \in T$.

If $\phi(\Delta)$ is of dimension 1, the inverses of 1-simplexes of T_2 are triangular or quadrilateral disks as shown in Figure II.2.A and the four-sided subdisks are subdivided as suggested by the dotted lines. $\square$

THEOREM II.2.B. *If (X, T) is a compact 2-complex and f is a linear map of (X, T) into R^2, then there is a triangulation T_2 of R^2 and a subdivision T_1 of T such that f takes each simplex of T_1 linearly onto a simplex of T_2.*

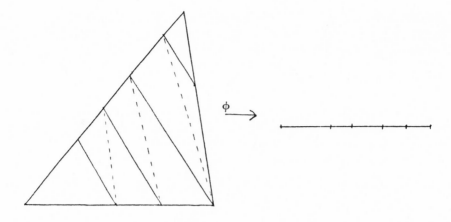

Figure II.2.A

PROOF. Let T_2 be a triangulation of R^2 such that the image under f of each simplex of T is the union of simplexes of T_2. Any refinement of T_2 would serve for the T_2 of the statement of the theorem but we use T_2 itself.

We get T_1 by subdividing each simplex Δ of T. We must not subdivide elements of T too much lest their images be proper subsets of elements of T_2. If $f(\Delta)$ is a vertex of T_2, $\Delta \in T_1$. If Δ^1 is a 1-simplex and f is a homeomorphism on Δ^1, f^{-1} on the triangulation T_2 of $f(\Delta^1)$ provides a subdivision of Δ^1. If Δ^2 is a 2-simplex of T, Theorem II.2.A provides a subdivision of Δ^2.   □

EXAMPLE II.2.C. Theorem II.2.B is false if one leaves out the hypothesis that X is compact. Consider the 1-dimensional case where X is the ray $[0, \infty)$ and T is a triangulation of X whose vertices are integral points. Let f take the segment $[i - 1, i]$ linearly onto the segment $[1/i, 1/(i + 1)]$. There is no subdivision T_1 of T and triangulation T_2 of R^2 such that f takes each simplex of T_1 *onto* a simplex of T_2.

THEOREM II.2.D. *If f is a linear map of a 2-complex (X_1, T_1) into a 2-complex (X_2, T_2) and S_2 is a subdivision of T_2, then there is a subdivision S_1 of T_1 so that f takes (X_1, S_1) linearly into (X_2, S_2).*

PROOF. We explain how to subdivide T_1 to get S_1. We do not subdivide a simplex Δ of T_1 if $f(\Delta)$ lies in an element of S_2.

If Δ^1 is a 1-simplex of T_1 and f is a homeomorphism on Δ^1, we get a description of S_1 on Δ^1 by subdividing Δ^1 so that the image under f of each of the pieces lies in an element of S_2. Similarly, if Δ^2 is a 2-simplex of T_1 and $f(\Delta^2)$ is of dimension 1, we have a situation like that depicted in Figure II.2.A and can subdivide Δ^2 in a manner as suggested there. If f is a homeomorphism on 2-simplex Δ^2 of T, we get a triangulation T' of $f(\Delta^2)$ such that if σ^1 is a 1-simplex of S_2 intersecting $f(\Delta^2)$, $\sigma^1 \cap f(\Delta^2) \in T'$. This is done without putting unnecessary vertices on Bd $f(\Delta^2)$. The elements of S_1 in Δ^2 are the inverses of the elements of T'.   □

THEOREM II.2.E. *If* (X_1, T_1), (X_2, T_2), (X_3, T_3) *are 2-complexes,* f_1 *is a PL map from* (X_1, T_1) *into* (X_2, T_2), *and* f_2 *is a PL map from* (X_2, T_2) *into* (X_3, T_3) *then* $f_2 f_1$ *is a PL map of* (X_1, T_1) *into* (X_3, T_3).

PROOF. Since f_2 is PL there is a refinement T_2' of T_2 such that f_2: $(X_2, T_2') \rightarrow (X_3, T_3)$ is linear. Also there is a refinement T_1' of T_1, such that f_1: $(X_1, T_1') \rightarrow (X_2, T_2)$ is linear. Since T_2' is a refinement of T_2, we find from Theorem II.2.D that there is a refinement T_1'' of T_1 such that f_1: $(X_1, T_1'') \rightarrow (X_2, T_2')$ is linear. Hence $f_2 f_1$ takes each simplex of T_1'' into a simplex of T_3.

Since the composition of linear maps is linear, $f_2 f_1$ is. □

II.3. Pushes. We shall describe a special useful kind of homeomorphism of R^2 onto itself called a *push* in R^2. Suppose v_0 is an arbitrary vertex of an arbitrary triangulation T of R^2. The *star of* v_0 (denoted by Star(v_0, T) or merely by Star v_0 if no confusion results) is the union of the simplexes of T containing v_0. The *link of* v_0 (denoted by Link v_0) is the union of all 1-simplexes $v_i v_j$ of T such that $v_0 v_i v_j$ is a 2-simplex of T.

A set X can be *starred* from a point $x \in X$ if each closed ray from x intersects X in a connected set. Note that the PL disk Star v_0 can be starred from v_0, and each ray from v_0 intersects the polygon Link $v_0 = \text{Bd Star } v_0$ in just one point.

Suppose p, $p' \in R^2$ and T, T' are rectilinear triangulations of R^2 such that Star p in T is the same set as Star p' in T'. The homeomorphism of R^2 onto itself that is fixed outside Star p, takes p to p', and takes the simplexes of T in Star p linearly onto simplexes of T' is called a *push*. Usually we regard the push of p to p' as a function of p, p' but not necessarily of T, T'. There are many pushes of p to p' depending on T's used. We think of a push as the end of a 1-parameter family of linear homeomorphisms of (R^2, T), the tth of which sends p to $(1 - t)p + tp'$.

A map h of a set X onto itself has *compact support* if X contains a compact set C such that h is the identity except possibly on C.

Note that a push in R^2 is a PL homeomorphism with compact support.

THEOREM II.3.A. *The composition of a finite number of pushes in* R^2 *is a PL homeomorphism with compact support.* □

Two complexes (X_1, T_1), (X_2, T_2) in R^2 are in *general position* there if the 0-skeletons of neither intersects either the 0-skelton or 1-skelton of the other. If a 1-simplex of T_1 intersects a 1-simplex of T_2, they cross. Pushes provide a convenient way to shift one complex so that it becomes in general position with respect to another. Additional consideration of general position is found in §IV.7. For example, (X_1, T_1), (X_2, T_2) are in general position in R^3 if no vertex of T_i ($i = 1, 2$) lies on a vertex, 1-simplex, or 2-simplex of T_j ($j \neq i$), and no 1-simplex of T_1 intersects any 1-simplex of T_2.

A finite complex (M, T) is called a compact 2-*manifold-with-boundary* if the star of each vertex of T is a disk.

THEOREM II.3.B. *Suppose in R^3 that (M_1, T_1), (M_2, T_2) are compact 2-manifolds-with-boundaries which are in general position. Then each component of $M_1 \cap M_2$ is either an arc or a simple closed curve.* $\square$

THEOREM II.3.C. *Suppose f_1, f_2 are linear maps of compact 2-manifolds-with-boundaries (M_1, T_1), (M_2, T_2) into R^3 such that each f_i is a homeomorphism and $f_1(T_1^i) \cap f_2(T_2^j) = \varnothing$ if $i + j < 3$. Then each component of $f_1^{-1}(f_1(M_1) \cap f_2(M_2))$ is either an arc or a simple closed curve.* $\square$

If a set X and its image $f(X)$ lies in the same metric space, we say that $d(f, \mathrm{Id}) \leqslant \varepsilon$ if for each $x \in X$, the distance $d(f(x), x)$ between a point x and its image $f(x)$ is no more than ε. We say that $d(f, \mathrm{Id}) < \varepsilon$ if there is a number δ such that $d(f, \mathrm{Id}) < \delta \leqslant \varepsilon$.

THEOREM II.3.D. *Suppose (X_1, T_1), (X_2, T_2) are two rectilinear complexes in R^2, and ε is a positive number. Then if X_1 is compact, there is a PL homeomorphism h: $R^2 \to R^2$ such that h is linear on (X_1, T_1), $d(h, \mathrm{Id}) < \varepsilon$, h has compact support, and (X_2, T_2), $(h(X_1), h(T_1))$ are in general position where $h(T_1)$ denotes the set of images under h of the simplexes of T_1.*

PROOF. Let $v_1, v_2, \ldots, v_n$ be the vertices of T_1 and T be a triangulation of R^2 such that $T_1 \subset T$. See Theorem I.1.B. We shall get a homeomorphism h by pushing v_1, then pushing $v_2, \ldots$, and finally pushing v_n. Each push will be less than ε.

To get the first push we get a point v_1' within ε of v_1 such that $\mathrm{Star}(v_1, T)$ is starred with respect to v_1' and v_1' does not lie on any straight line determined by pairs of vertices in $T_1 \cup T_2$. The push h from v_1 to v_1' changes the 1-simplexes of T_1 to 1-simplexes that are in general position with respect to (X_2, T_2). For simplicity we suppose h was the identity so as to be able to refer to T_1 instead of $h(T_1)$.

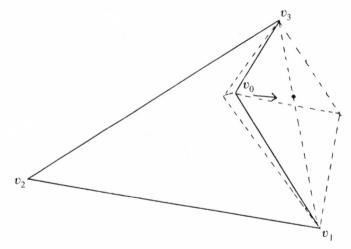

Figure II.3.E

The proof is continued by pushing v_2, then $v_3, \ldots,$ and finally v_n. □

It is to be noted that we could have dropped the condition that X_1 is compact if we had not required that h have compact support.

The next two exercises along with the theorem which follows are included as preliminaries to Chapter III.

EXERCISE II.3.E. If $v_0 v_1 v_2 v_3$ is a concave quadrilateral as shown in Figure II.3.E, there is a push h that sends $v_0 v_1 v_2 v_3$ to a triangle.

A suitable push sends v_0 to some point of Int $v_1 v_3 = (v_1 v_3)$. This exercise is included not because it is significant or interesting but rather because it is related to the important PL Schoenflies theorem discussed in the next chapter. The following exercise is another variation of the same idea.   □

EXERCISE II.3.F. If $v_1 v_2 v_3 v_4 v_5$ is a regular polygon as shown in Figure II.3.F, there is a push that sends $v_1 v_2 v_3 v_4 v_5$ to a quadrilateral.   □

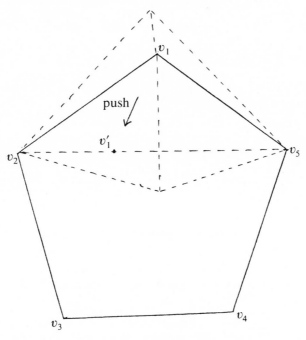

Figure II.3.F

By pushing the vertices of a triangle, one gets the following result which is used in the next chapter.

THEOREM II.3.G. *Suppose D is a triangular disk in R^2 and x, y, z are three noncollinear points of* Bd D, *then there are pushes h_1, h_2, h_3 on the vertices of D such that $h_3 h_2 h_1(D)$ is a triangular disk with vertices x, y, z.*   □

II.4. Isotopies. Suppose $h\colon X \times [0, 1] \to Y$ is continuous. We may denote $H(x \times t)$ by $H_t(x)$ and call H_t $(0 \leq t \leq 1)$ a 1-parameter family of maps of X into Y. A member of the family $\{H_t\}$ is a map H_t of X where $H_t(x) = H(x \times t)$.

We shall not try to define a distance between elements of $\{H_t\}$. Note that X need not be compact. We think of the family $\{H_t\}$ as being continuous since H: $X \times [0, 1] \to Y$ is continuous.

An *isotopy* of a space X into a space Y is a 1-parameter family of homeomorphisms H_t $(0 \leqslant t \leqslant 1)$ of X into Y. It is required that $\{H_t\}$ be simultaneously continuous in t and X rather than that the homeomorphisms be near each other. For example, a rotation of R^2 about the origin generates an isotopy even though each pair of the homeomorphisms are infinitely far apart. (Some mathematicians prefer to regard an isotopy as a map H of $X \times [0, 1]$ into Y such that H restricted to each $X \times t$ is a homeomorphism. This concept is equivalent to the one we use.) In many cases we will be considering the case where $Y = X$, each H_t is onto, and one of H_0, H_1 is the identity. One can think of such an isotopy as a motion of X onto itself using t as a time variable. A push is such an isotopy.

The following theorem generalizes to all dimensions.

THEOREM II.4.A. *If h is a homeomorphism of a disk D onto itself that is fixed on* Bd D, *then there is an isotopy H_t $(0 \leqslant t \leqslant 1)$ of D onto itself such that $H_0 = h$, $H_1 = $ identity, and each H_t is fixed on* Bd D.

PROOF. There are several proofs of this result, but we think of the "meat grinder" approach suggested by the idea that when material is forced through a sieve or meat grinder, it comes out in smooth threads. Suppose that one starts with D as distorted by h and straightens it out by shrinking Bd D toward the center of D. See Figure II.4.A, where D is regarded as a rectangle, and we show three views of it. If at time t, Bd D has been squeezed to the boundary of the smaller rectangle shown in the upper left part of Figure II.4.A, then H_t is the identity in the part of D outside this rectangle. To describe H_t on the small rectangle, one might use a linear homeomorphism to expand the smaller rectangle and make it fit onto the larger, then use $H_0 = h$ to send this larger rectangle onto the upper right version of D, and finally use a linear homeomorphism to send the rectangle at the upper right of Figure II.4.A onto the small rectangle in the lower center version of D. $\square$

A homeomorphism h on X is called an ε-homeomorphism if $d(x, h(x)) \leqslant \varepsilon$ for each $x \in X$. The following theorem generalizes easily to all dimensions.

THEOREM II.4.B. *Suppose $\varepsilon > 0$ and h is an ε-homeomorphism of R^2 onto itself. Then there is an isotopy H_t $(0 \leqslant t \leqslant 1)$ of R^2 onto itself such that $H_1 = h$, $H_0 = $ identity, and each H_t is an ε-homeomorphism.*

Recall that for each point $p = (x, y) \in R^2$, tp is the point with coordinates (tx, ty). Let g_t be the homeomorphism of R^2 onto itself that sends each point p to tp. In proving Theorem II.4.B, J. M. Kister [$\mathbf{K_3}$] made use of the function $g_t h g_{1/t}$. Note that if $0 < t < 1$, then $g_t h g_{1/t}(p) = H_t$ $(0 < t < 1)$ is obtained by pushing p away from the origin, then moving it by h, and finally pulling it toward the origin.

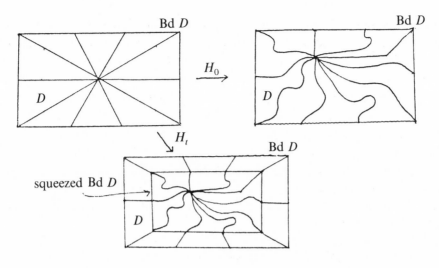

Figure II.4.A

One uses the hypothesis that h is an ε-homeomorphism to show $H_0 = \lim_{t \to 0} g_t h g_{1/t}$ is the identity. $\square$

II.5. Meshing triangulations. Suppose as shown in Figure II.5 that U_1, U_2 are open sets with curvilinear triangulations T_1, T_2 which do not agree on $U_1 \cap U_2$. We seek a triangulation T of $U_1 \cup U_2$ that is compatible with T_1 on $U_1 - \overline{U}_2$ and with T_2 on $U_2 - \overline{U}_1$. We say that two triangulations S_1, S_2 of X_1, X_2 are *locally compatible* at a point $p \in X_1 \cap X_2$ if there are subdivisions S_1', S_2' of S_1, S_2 such that the identity map takes the union of the simplexes of S_1' containing p linearly onto the union of the simplexes of S_2' containing p. Also, S_1 and S_2 are *compatible* on $X_1 \cap X_2$ if there is a triangulation T of $X_1 \cap X_2$ which is subdivision of both S_1 and S_2.

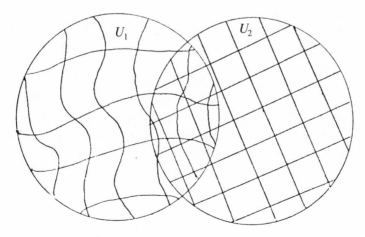

Figure II.5

If T is a rectilinear triangulation of an open subset of R^2, then any 1-simplex of T is an edge of two and only two 2-simplexes of T. It follows from Chapter III that curvilinear triangulations of open subsets of R^2 also have this property, and we tacitly assume it in the next theorem, a variation of which can be used to show that the union of certain triangulated sets can be triangulated.

THEOREM. II.5. *Suppose*

U_1, U_2 *are two open subsets of* R^2,

$T_i\ (i = 1, 2)$ *is a curvilinear triangulation of* U_i,

B *is the union of 1-simplexes* Δ^1 *of* T_1 *such that one but not both of the 2-simplexes of* T_1 *with* Δ^1 *as an edge lies in* U_2,

h *is a homeomorphism of* R^2 *onto itself that is fixed off* $U_1 \cap U_2$ *and takes each 1-simplex of* T_1 *in* B *onto a 1-simplex of* T_2.

Then there is a curvilinear triangulation T *of* $U_1 \cup U_2$ *such that*

T *is compatible with* T_1 *on* $U_1 - \overline{U}_2$ *and with* T_2 *on* $U_2 - \overline{U}_1$.

PROOF. Let M be union of 2-simplexes of T_1 that intersect $U_1 - U_2$. Note that $B = \mathrm{Bd}\ M$ in U_1. If Δ is a simplex of T_1 in M, $h(\Delta)$ is an element of T. If Δ is a simplex of T_2 not in $h(M)$, $h(M) \cap \mathrm{Int}\ \Delta = \varnothing$. Each simplex of T_2 not in $h(M)$ is an element of T.

It remains to assign $U_1 \cup U_2$ a metric so that each simplex of T is isometric to a triangular disk. We shall do this so that each is isometric to a unit equalateral triangular disk D_0^2. It is easy to get a global metric if we metrize each simplex of T so we concentrate on the elements of T.

If Δ^2 is a simplex of T_1 in M, there is a linear homeomorphism g of D_0^2 onto Δ^2 under T_1, and we pick a metric for $h(\Delta^2)$ of T by deciding that $hg: D_0^2 \to h(\Delta^2)$ is an isometry. As a first approximation to metrics on other 2-simplexes of T, we pick a metric for a 2-simplex $\Delta^2 \in T$ in $U_2 - h(\mathrm{Int}\ M)$ by letting g be a linear homeomorphism of D_0^2 onto Δ^2 under T_2 and deciding that g is an isometry. The defect in this metric is that if Δ^1 is a 1-simplex of T_1 in B, we have assigned two metrices to $h(\Delta^1)$. The situation is remedied by considering a 2-simplex Δ^2 of T_2 that does not lie in $h(M)$ but contains an edge Δ^1 in $h(B)$ and modifying the metric on Δ^2. Let f be a homeomorphism Δ^1 onto itself that is fixed on Bd Δ^1 and is an isometry in taking the second metric (one associated with T_2) into the first. Let p be a point in Int Δ^2 so near Δ^1 that the cone from p over Δ^1 lies in $U_1 \cup U_2$ and F be the homeomorphism of Δ^2 onto itself that is the identity off of the cone and agrees with f on Δ^1. Then Δ^2 is metrized so that $Fg: D_0^2 \to \Delta^2$ is an isometry. This modification is done for each 1-simplex in $h(B)$. Now the metrics for two 2-simplexes of T with a common edge do agree on the common edge. $\square$

CHAPTER III

THE SCHOENFLIES THEOREM

The methods used in this chapter are special for the plane and most are not known to generalize to all higher dimensions. Some are even false there.

The Schoenflies theorem (Theorem II.6.C) has played an important role in Euclidean topology. It shows that if J is any simple closed curve in R^2 whatsoever, then there is a homeomorphism of R^2 onto itself that takes J onto a circle. It is used to discover other useful properties—for example, it is used to show that the Euler characteristic of a 2-sphere is 2. It can be used to show that the cranky neighbor problem has no solution.

Early proofs of the Schoenflies theorem were complicated. Some mathematicians may not have sought proofs since they felt that the result was intuitively obvious and hence near axiomatic. It may have come as a surprise to some to learn that the corresponding theorem in the next dimension is false. Chapter IV describes wild 2-spheres in R^3.

A proof of the Schoenflies theorem for the plane has been known since 1906. That each simple closed curve separated R^2 into precisely two pieces was already known. Carathéodory is credited with a proof of the Schoenflies theorem from the realm of complex numbers. He used conformal mappings to show that if J is a simple closed curve in R^2, then there is a homeomorphism of the unit disk $|z| \leq 1$ onto the union of J and the bounded component of $R^2 - J$. This treatment had the advantage that the homeomorphism preserved right angles for $|z| < 1$, but it had the disadvantage that it gave an existence proof rather than a constructive one and left many students in the dark as to where points went. Taking a unit disk into the disk bounded by J gives one form of the Schoenflies theorem (like Theorem III.6.A), but this form is easily translated into others as shown by the proofs of Theorems III.6.B and III.6.C.

We now have at hand (see Cairns [$\mathbf{C_1}$] and Moise [$\mathbf{M_{12}}$]) several good proofs of the Schoenflies theorem. One of the easiest proofs makes use of the PL Schoenflies theorem. This PL version has merits of its own so we use that approach. We prove the PL Schoenflies theorem in §III.1, note some applications of it in §§III.2, III.3, and III.4, and then treat the Schoenflies theorem in III.5 and III.6.

19

III.1. PL Schoenflies theorem. A *polygon* is a simple closed curve that is the sum of a finite number of 1-simplexes. The places where it bends are called *vertices*, and a 1-simplex joining adjacent vertices is called an *edge* or *side*.

THEOREM III.1.A (PL SCHOENFLIES THEOREM). *For each polygon P in the plane R^2 there is a finite collection of pushes whose composition sends P to a triangle.*

PROOF. The proof is by induction on the number of sides of P. It is clear that the result holds if P has 3 sides. Let n be an integer larger than 3 such that if $3 \leq k \leq n - 1$, Theorem III.1.A holds for any polygon with k sides. We then finish the proof by showing that Theorem III.1.A holds for any polygon with n sides.

Assume P has n sides. It follows from Theorem III.1.D (which we assume for now) that there is a nonadjacent pair (say p, q) of vertices of P such that the straight 1-simplex pq intersects P only in its ends. Then P is the union of two PL arcs pxq, pyq. The points x, y are points of the open arcs (pxq), (pyq). They need not be vertices and are used merely to distinguish between the three arcs pq, pxq, pyq from p to q. We call $pxq \cup pyq \cup pq$ a θ-curve. The polygons $pxq \cup pq$ and $pyq \cup pq$ each has fewer than n edges so Theorem III.1.A holds for each of them.

Let h_1, h_2 be finite compositions of pushes in R^2 that take $pxq \cup pq$, $pyq \cup pq$, respectively, to triangles. We do not suppose the images of p, x, y, q are vertices. Where does $h_1(\text{Int } pyq)$ lie? There are two cases according as to whether it lies outside or inside the triangular disk bounded by $h_1(pxq \cup pq)$.

If $h_1(\text{Int } pyq)$ does not intersect the 2-simplex bounded by $h_1(pxq \cup pq)$ we find that there is a composition h_3 of three pushes so that $h_3h_1(pxq \cup pq)$ is a triangle with vertices $h_3h_1(p)$, $h_3h_1(q)$, $h_3h_1(x)$. Also, $h_3h_1(\text{Int } pyq)$ does not lie in this triangle. As noted in Figure III.1.A there is a push h_4 at $h_3h_1(x)$ such that h_4 is fixed on $h_3h_1(pyq)$ but $h_4h_3h_1(pxq)$ is the 1-simplex $h_3h_1(pq)$. Check that $h_1^{-1}h_3^{-1}h_4h_3h_1$ takes $pxq \cup pyq$ to $pq \cup pyq$ and $h_2h_1^{-1}h_3^{-1}h_4h_3h_1$ takes P to a triangle.

If $h_1(\text{Int } pyq)$ intersects the 2-simplex bounded by $h_1(pxq \cup pq)$, we interchange the roles of h_1 and h_2 and proceed as before. In this case $h_1(pyq)$ lies in the 2-simplex bounded by $h_1(pxq \cup pq)$ and $h_1(x)$ is accessible from the unbounded components of $h_1(P \cup pq)$. Then $h_2(x)$ does not lie in the triangular disk bounded by $h_2(pyq \cup pq)$ and $h_2(\text{Int}(pxq))$ does not intersect the disk. $\square$

We could have broken our argument for Theorem III.1.A with the following halfway theorem. Since we have achieved the whole, we will not belabor the half.

THEOREM III.1.B. *Suppose Q is a PL θ curve in R^2. If Theorem III.1.A holds for two of the simple closed curves in θ, it holds for the third.* $\square$

We now express the PL Schoenflies theorem in a familiar form which follows from Theorem III.1.A.

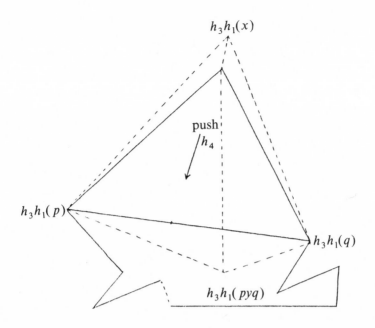

$$h_3h_1(x)$$

push $/h_4$

$h_3h_1(p)$

$h_3h_1(q)$

$h_3h_1(pyq)$

FIGURE III.1.A

THEOREM III.1.A′ (PL SCHOENFLIES THEOREM). *For each pair of polygons P_1, P_2 in R^2 there is a PL homeomorphism with compact support of R^2 onto itself that takes P_1 to P_2.* □

There are generalizations of the above theorem saying how we can keep the required PL homeomorphism fixed on certain sets or make the PL homeomorphism agree with certain preassigned PL homeomorphisms of P_1 onto P_2. However, we will not pursue these.

By the PL Schoenflies theorem we can speak not only of the interior and exterior of a triangular polygon but also of any polygon P in R^2. Int P and Ext P are the bounded and unbounded components respectively of $R^2 - P$.

THEOREM III.1.C. *For each pair of polygons P_1, P_2 in R^2, each homeomorphism h of P_1 onto P_2 can be extended to a homeomorphism of $P_1 \cup$ Int P_1 onto $P_2 \cup$ Int P_2.*

PROOF. Let g_i be a homeomophism of $P_i \cup$ Int P_i onto a triangular disk D as guaranteed by Theorem III.1.A. Select a point p_0 of Int D and let $h': D \to D$ be the homeomorphism such that if $y \in$ Bd D, h' takes the 1-simplex $p_0 y$ linearly onto the 1-simplex from p_0 to $g_2hg_1^{-1}(y)$. Then $g_2^{-1}h'g_1$ is a suitable extension of h. □

THEOREM III.1.D. *If P is a polygon with more than three vertices in R^2, some pair of vertices of P can be joined by a 1-simplex that intersects P at only these two points.*

PROOF. Let v_1, v_2, v_3 be three consecutive vertices of P. If $v_1 v_3$ does not serve for the required 1-simplex, $(v_1 v_3)$ intersects P. If Int $v_1 v_2 v_3$ contains a vertex of P, let v be the farthest such vertex from the line through v_1, v_3 and note that $v_2 v$ will serve. If $(v_1 v_3)$ intersects P but no vertex of P lies in Int $v_1 v_2 v_3$, then $(v_1 v_3)$ contains a vertex v of P and $v_2 v$ will serve. $\square$

ALTERNATE PROOF. Those acquainted with convex sets and lines of support may prefer the following proof. If P bounds a convex disk, any nonadjacent pair of vertices will serve. If it does not bound a convex disk, some line of support intersects P in a nonconnected set. This line of support contains a 1-simplex verifying the theorem. $\square$

III.2. Triangulating PL disks. This section gives some applications of the PL Schoenflies theorem.

A PL *disk* is the homeomorphic PL image of a 2-simplex. A polygon P in R^2 bounds a disk D and we write Bd $D = P$. Sometimes a PL disk is called a *polygonal disk*. If (X, T) is a complex, we say that X is PL. Here the symbol PL indicates polygonal or polyhedral. Recall from §II.2 that it meant piecewise linear when referring to a map.

THEOREM III.2.A. *If Q is a PL θ-curve in R^2, the three disks D_1, D_2, D_3 bounded by the three polygons in Q can be ordered D_1, D_2, D_3 so that $D_3 = D_1 \cup D_2$ and $D_1 \cap D_2$ is an arc in Q.* $\square$

We plan to get an efficient triangulation of a polygonal disk D in R^2. To this end we consider certain spanning arcs of D. An arc ab in D *spans* D if it intersects Bd D only in the arc's ends. A spanning arc ab is a *proper spanning segment* if it is straight and its ends are vertices of Bd D.

THEOREM III.2.B. *Each PL disk D in R^2 with more than 3 sides has a proper spanning segment.*

PROOF. Let v_1, v_2, v_3 be three consecutive vertices of Bd D. The proof breaks into two cases according as to whether or not the part of Int $v_1 v_2 v_3$ near v_2 lies in D.

If it lies in D, we use the technique of the proof of Theorem III.1.D to obtain a proper spanning segment.

If it does not lie in D, we extend a ray from v_2 in a direction that starts the ray into Int D. The open ray first hits P at a point x. If x is a vertex of Bd D, $v_2 x$ is a proper spanning segment. If it is not a vertex, x lies on a side ab of Bd D—pick a to be the one of these such that neither v_1 or v_3 lies in the 2-simplex $a x v_2$. See Figure III.2.B. Suppose $v_2 a$ is not a proper spanning segment. Let y be the first point of xa in the order from x to a such that $v_2 y$ intersects Bd D in more than two points. A subarc of ay is a spanning segment from v_2.

Note that we have not shown that there is a proper spanning segment from v_2. However there is if $v_1 v_3$ is not a proper spanning segment. $\square$

Repeated applications of Theorem III.2.B gives the following result.

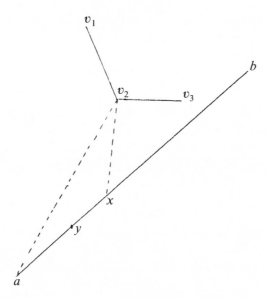

FIGURE III.2.B

THEOREM III.2.C. *If P is a polygon in R^2, then there is a triangulation T of the polygonal disk bounded by P such that each 0-simplex of T is a vertex of P and each 1-simplex of T is either an edge of P or a proper spanning segment of P.* □

COROLLARY III.2.D. *If P is a polygon in R^2, then there is a triangulation T of the convex hull of P such that each vertex of T is a vertex of P and each edge of P is a 1-simplex of T.* □

III.3. **Skew curves.** We give some further applications of the PL Schoenflies theorem. A *finite graph* is the union of the finite number of arcs such that if two of them intersect, their intersection is an end point of each. We shall be concerned with restrictions on the embeddability of certain finite graphs in R^2.

THEOREM III.3.A. *Suppose a_1, b_1, a_2, b_2 are points as ordered on a polygon P in R^2 and a_1a_2, b_1b_2 are disjoint PL arcs which intersect P only at their ends. Then (a_1a_2), (b_1b_2) lie in different components of $R^2 - P$.*

PROOF. Apply Theorem II.2.A to $P \cup a_1a_2$ as shown in Figure III.3.A. Note that (b_1b_2) lies in one of Int D_1, Int D_2, Ext D_3.

If a_1a_2 is the middle arc $D_1 \cap D_2$, b_1 and b_2 are on outside arcs and $(b_1b_2) \subset$ Ext D_3. In this case P separates (a_1a_2) from (b_1b_2).

If a_1a_2 is the top arc, $(b_1b_2) \subset$ Int D_2 and P separates (a_1a_2) from (b_1b_2).

If a_1a_2 is the bottom arc, $(b_1b_2) \subset$ Int D_1 and P separates (a_1a_2) from (b_1b_2). □

We shall show that certain finite graphs cannot be embedded in R^2. In view of our reliance on the PL Schoenflies theorem we show that if certain graphs could be embedded, topologically equivalent PL graphs could be embedded also.

THEOREM III.3.B. *If p, q are two points of a connected open set U of R^2, there is a PL arc in U from p to q.* $\square$

THEOREM III.3.C. *For each finite graph G in R^2 and each $\varepsilon > 0$ there is a homeomorphism h: $G \to R^2$ such that $d(h, \text{Id}) < \varepsilon$ and $h(G)$ is PL.*

PROOF. Subdivide arcs making up G so that G is the union of very small arcs $A_1, A_2, \ldots, A_n$ such that if two of the A's intersect, their intersection is an end of each. Put very very small round disks about the ends of the A's and replace each A_i by the union of a radial arc in each of the two disks about the ends and a PL arc near A_i joining the radial ends. $\square$

A *skew curve of type* 1 is a set homeomorphic to the 1-complex shown in Figure III.3.D. This 1-complex has six vertices $a_1, a_2, a_3, b_1, b_2, b_3$ and nine arcs $a_1b_1, a_1b_2, \ldots, a_3b_3$ as shown. The arcs are not assumed to be straight. The curve is sometimes called the cranky neighbor graph (or the utility example) because of the question as to whether or not three neighborhoods a_1, a_2, a_3 can run utility lines to three stations b_1, b_2, b_3 so that their utility lines will not cross.

THEOREM. III.3.D. *No skew curve of type* 1 *lies in the plane.*

PROOF. In view of Theorem III.3.C we suppose that the skew curve of type 1 under consideration is PL. Suppose $a_1b_1a_2b_2$ is the P of Theorem III.3.A and

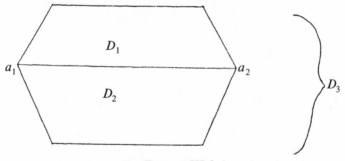

FIGURE III.3.A

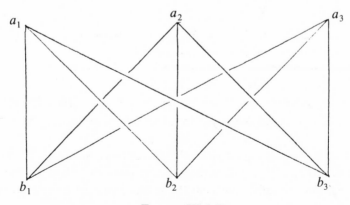

FIGURE III.3.D

$(a_1b_3a_2)$, $(b_1a_3b_2)$ are the open arcs (a_1a_2), (b_1b_2) of that theorem. If the skew curve lies in R^2, a_3b_3 shows that P does not separate $(a_1b_3a_2)$ from $(b_1a_3b_3)$. This violates Theorem III.3.A. □

A *skew curve of type* 2 is a set homeomorphic to the complete graph on five vertices as shown in Figure III.3.E. Kuratowski showed that a finite graph could be embedded in the plane if it did not contain either a skew curve of type 1 or one of type 2.

THEOREM III.3.E. *No skew curve of type 2 lies in the plane.*

PROOF. Again we assume the graph under consideration in PL and assume it lies in the plane. We let $a_1a_2a_3a_4a_5$ be the P of Theorem III.3.A and denote the components of $R^2 - P$ by U and V where $(a_2a_5) \subset U$. It follows from Theorem III.3.A that (a_1a_3) and (a_1a_4) lie in V while (a_2a_4), (a_3a_5) lie in U. This contradicts Theorem III.3.A which states that (a_2a_4) and (a_3a_5) lie in different components of $R^2 - P$. The contradiction arose from the faulty assumption that some skew curve of type 2 lies in R^2. □

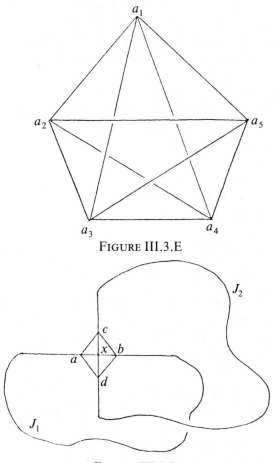

FIGURE III.3.E

FIGURE III.3.F

III.4. No arc separates R^2. It is sometimes useful to partition a set. If T is a triangulation of the boundary of a tetrahedron, the collection of interiors of 2-simplexes of T is a partitioning of the boundary. A *partitioning* of a set X is a finite collection of mutually disjoint open subsets of X whose union is dense in X.

A partitioning is a *brick partitioning* if each of its elements is the interior of its closure, is uniformly locally connected, and if the interior of the closure of the union of each two elements is also uniformly locally connected. Here *uniformly locally connected* means 0-ulc as defined in §5 of Chapter VIII. One can get a brick partitioning of the boundary of a tetrahedron by letting T be a triangulation of the boundary, T' be the barycentric subdivision of T, and letting the elements of the partitioning be the open stars in T' of the vertices of T.

Bing used brick partitionings in [$\mathbf{B_6}$] to show that each continuous curve has a convex metric. He used them in [$\mathbf{B_3}$] to improve the Kline sphere characterization. We need only a mild form of it in the proof of Theorem III.4.A, and there we may use as elements of the partitionings the interiors of rectangles stacked so that the interior of the closure of the union of any two of them is uniformly locally connected. The boundary of the closure of the union of any subset of them is the union of a finite number of mutually disjoint polygons.

There are many proofs of the result that no arc separates R^2. The one we use here makes a pleasant application of the PL Schoenflies theorem. A variation of the methods used here can be used to get related results in higher dimensions. See Chapter VIII for some of these extended results.

THEOREM III.4.A (JANISZEWSKI). *Suppose that in* R^2
A, B are two closed sets,
A is compact,
p, q are two points of $R^2 - (A \cup B)$,
paq is a polygonal arc from p to q in $R^2 - B$,
pbq is a polygonal arc from p to q in $R^2 - A$, *and*
$A \cap B$ is connected or null.
Then there is a polygonal arc pxq from p to q such that

$$pxq \subset R^2 - (A \cup B).$$

PROOF. We suppose that the theorem is false, that $paq \cap pbq$ is finite, and that r is the first point of $paq \cap pbq$ in the order from p to q on paq such that p and r belong to different components of $R^2 - (A \cup B)$. For convenience we suppose that $r = q$, that $paq \cup pbq$ bounds a rectangular disk D as shown in Figure III.4.A, and that $B \cap qa = \varnothing$.

First we consider the case where $A \cap B \cap D = \varnothing$. Let ε be a positive number less than $d((A \cup pa), (B \cup bq))$ and $\{D_i\}$ be a brick partitioning of D into rectangular disks as shown in Figure III.4.A so that each D_i is of diameter less than ε and each component of the boundary of the union of a finite number of the D_i's is a polygon. Let X be the union of the D_i's intersecting $A \cup pa$ and P be the component of Bd X containing pa. Then $(P - (pa)) \cup qa$ contains an arc in

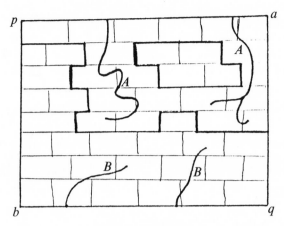

FIGURE III.4.A

$D - (A \cup B)$ joining p and q which shows the assumption that the theorem is false is wrong.

The treatment in case $A \cap B$ misses $R^2 - D$ is similar except that we partition $R^2 - \text{Int } D$ instead of D. There are infinitely many D_i's but only a finite number intersect the compact set A. □

THEOREM III.4.B. *No arc* (*topological*) *in R^2 separates R^2.*

PROOF. Suppose an arc ab in R^2 separates p from q. Let $a = a_1, a_2, \ldots, a_n = b$ be points on ab so close together than no $a_i a_{i+1}$ separates p from q. Iterated use of Theorem III.4.A shows that $a_1 a_3, a_1 a_4, \ldots, a_1 a_n = ab$ does not either. □

III.5. Jordan-Brouwer theorem. The Jordan curve theorem says that each simple closed curve J in R^2 separates R^2 and each component of $R^2 - J$ has J for a boundary. The Jordan-Brouwer theorem puts on the additional conclusion that $R^2 - J$ has exactly two components. The Jordan-Brouwer theorem plays an important role in plane topology. R. L. Moore [M_{14}] used it as one of the axioms in one of this axiomatic treatments of plane topology.

The Jordan-Brouwer theorem is weaker than the Schoenflies theorem but unlike it, remains true for $(n - 1)$-spheres in R^n. We use a proof based on applications of the PL Schoenflies theorem because the proof is easy. Later we will shift to a more complicated proof for higher dimensions.

A simple closed curve in R^2 is a *semipolygon* if it contains a straight arc. The open straight are is called a *flat spot*.

THEOREM III.5.A. *A semipolygon J in R^2 separates R^2.*

PROOF. Let axb be a straight arc in J and cxd be a short straight arc intersecting J only at x. Then J separates c from d or else there is a simple closed curve J_2 containing cxd, intersecting J only at $\{x\}$, and violating Theorem III.3.F. □

The following is an important application of Theorem III.4.B.

THEOREM III.5.B. *If S is a semipolygon in R^2 and U is component of $R^2 - S$, then each point S is a limit point of U.* □

A set X is *accessible* from a set Y at a point $p \in X$ if there is an arc pq in $X \cup Y$ such that pq lies in $Y \cup \{p\}$. If pq is straight, we say p is accessible from Y along a straight arc. The following is an important application of Theorem III.4.B.

THEOREM III.5.C. *If J is a simple closed curve in R^2 and U is a component of $R^2 - J$, then there is a dense set X of points of J such that if $p \in X$, J is accessible at p from U along a straight arc.*

PROOF. We show that for each $q \in J$ and each neighborhood N of q there is a point $p \in N \cap J$ such that p is accessible from U along a straight arc. Let ab be an arc in $J - \{q\}$ which contains $J - N$. It follows from Theorem III.4.B that there is a PL arc cq in $R^2 - ab$ from a point c of U to q. The first point of J on cq in order from c is a suitable p. $\square$

Theorem III.5.C implies the following.

THEOREM III.5.D. *If J is a simple closed curve in R^2 and U is a component of $R^2 - J$, then each point of J is a limit point of U.* $\square$

THEOREM III.5.E. *Each simple closed curve J in R^2 separates R^2.*

PROOF. Let pq be a straight arc that intersects J only at its ends. Denote the two arcs in J from p to q by paq, pbq. Let D be a round disk about a which misses $pbq \cup pq$. It follows from Theorems III.5.A and III.5.C that there are points r, s in D which belong to different components of $R^2 - (paq \cup pq)$. We shall show that J separates r from s.

Assume J does not separate r from s and rcs is a PL arc in $R^2 - J$ from r to s. We adjust rcs to reduce its intersection with pq. With no loss of generality we suppose $c \in pq$ and rcs intersects pq only at a flat spot of rcs. Let rs be a straight arc in D and J_2 be a simple closed curve in $rs \cup rcs$ that contains c.

Then J_2 separates p from q because of the crossing of pq and rcs. However it does not separate them because of pbq. The assumption that J did not separate r from s led to this contradiction. $\square$

THEOREM III.5.F. *If J is a simple closed curve in R^2, $R^2 - J$ does not have more than two components.*

PROOF. Suppose $R^2 - J$ has 3 components U_1, U_2, U_3. Let $a_1, b_1, a_2, b_3, a_3, b_2,$ a_1 be six points on J ordered as listed. It follows from Theorems III.3.B and III.5.C that there is an open arc from a point of J near a_1 to a point of J near b_3 in U_1, an open arc from a point of J near b_1 to a one near a_3 in U_2, and an open arc from a point of J near a_2 to one near b_2 in U_3. This contradicts Theorem III.3.D. $\square$

We combine Theorems III.5.C, III.5.D, III.5.E, and III.5.F to get the following.

THEOREM III.5.G (JORDAN-BROUWER THEOREM). *If J is a simple closed curve in R^2, then $R^2 - J$ has exactly two components and is the boundary of each.* $\square$

After a treatment of the Jordan-Brouwer theorem we are in position to speak of the interior of any simple closed curve whatsoever in R^2. Theorem III.2.A can be restated as

THEOREM III.5.H. *If Q is a θ-curve in R^2, the three simple closed curves in Q can be ordered J_1, J_2, J_3 so that $J_3 \cup \text{Int } J_3 = (J_1 \cup \text{Int } J_1) \cup (J_2 \cup \text{Int } J_2)$ and $(J_1 \cup \text{Int } J_1) \cap (J_2 \cup \text{Int } J_2)$ is an arc in Q.* $\square$

III.6. **Schoenflies theorem.** We now come to the crux of the proof of the Schoenflies theorem. In a basic preliminary step we extend the homeomorphism h of a simple closed curve J in R^2 onto a triangular polygon to a homeomorphism of $J \cup \text{Int } J$ onto the triangular disk bounded by the polygon. The proof uses infinitely many steps. Figure III.6.A.a suggests a start. Theorems III.6.A, III.6.B, III.6.C are all versions of the Schoenflies theorem.

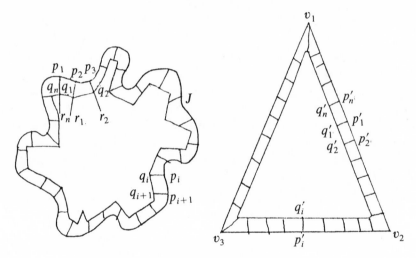

FIGURE III.6.A.a

THEOREM III.6.A. *Suppose in R^2 that J is a simple closed curve and D is a triangular disk. Any homeomorphism h of J onto $\text{Bd } D$ can be extended to take $J \cup \text{Int } J$ onto D.*

PROOF. We use a sequential description of the extension of h and to emphasize that we will be making repeated uses of the same technique, we let $\varepsilon_1, \varepsilon_2, \ldots$ be a sequence of positive numbers converging to 0. The harmonic sequence $1, \frac{1}{2}, \frac{1}{3}, \ldots$ would do, but the choice is unimportant.

Theorem III.5.C allows us to pick points $p_1, p_2, \ldots, p_n$ cyclically ordered on J so abundant and so close together that the arcs $p_i p_{i+1}$ on J and their images on $\text{Bd } D$ are small (have diameters less than $\varepsilon_1/3$) and the p_i's are accessible from $\text{Int } J$ by straight arcs. (We treat p_{n+1} as p_1.)

Let $p_i r_i$ be mutually disjoint short (diameters less than $\varepsilon_1/3$) straight arcs which lie, except for their end p_i, in $\text{Int } J$. Designate $h(p_i)$ on $\text{Bd } D$ by p_i' and let $p_i' q_i'$ be mutually disjoint short (diameters less than $\varepsilon_1/3$) straight arcs which lie except for p_i' in Int D.

It follows from repeated applications of Theorem III.6.D (which we will prove later) that there is a polygon $q_1q_2 \cdots q_n$ in Int J (the polygon may have more than n vertices) and straight arcs $p_1q_1, p_2q_2, \ldots, p_nq_n$ as shown in Figure III.6.A.a. Here each $p_iq_i \cup q_iq_{i+1} \cup q_{i+l}p_{i+1}$ lies in an $\varepsilon_1/3$ neighborhood of p_ip_{i+1}. The polygon $q_1q_2 \cdots q_n$ may be constructed as follows. Apply Theorem III.6.D directly to get an arc s_1s_2 in U such that s_1s_2 is close to p_1p_2 and intersects p_ir_i only at s_i ($i = 1, 2$). We suppose $p_1s_1 \cup s_1s_2 \cup s_2p_2$ misses superfluous p_jr_j's. Let O_2 be a neighborhood of p_2p_3 that misses $s_1s_2 \cup s_2r_2$ and superfluous p_jr_j's and apply Theorem III.6.D to get an arc q_2s_3 in $O_2 \cap U$ irreducible from p_2r_2 to p_3r_3. Note that q_2 is between s_2 and p_2. We continue building $q_3s_4, q_4s_5, \ldots, q_nq_1$. The $q_1q_2 \cdots q_n$ is a union of s_1s_2, (s_2p_2 of p_2r_2), q_2s_2, (s_3q_3 of p_3r_3), $\ldots, q_nq_1$, (q_1s_1 of p_1r_1).

We now extend h to take the p_iq_i's to the $p_i'q_i'$'s and the polygon $q_1q_2 \cdots q_n$ to $q_1'q_2' \cdots q_n'$. It is further extended to take the PL disk in Int J bounded by $q_1q_2 \cdots q_n$ onto the disk in Int D bounded by $p_1'q_2' \cdots q_n'$. This last extension is allowed by Theorem III.1.C.

We have now extended h to define it on all of $J \cup$ Int J except on the interiors of some small simple closed curves. The $\varepsilon_1/3$ restriction insures that these simple closed curves and their images in D have diameters less than ε_1. A typical simple closed curve is shown in Figure III.6.A.a as $p_iq_i \cup (q_iq_{i+1}$ of $q_1q_2 \cdots q_n) \cup q_{i+1}p_{i+1} \cup (p_{i+1}p_i$ of $J)$ and in Figure III.6.A.b as $J_i = p_ir_iq_i \cup q_ib_iq_{i+1} \cup q_{i+1}r_{i+1}p_{i+1} \cup p_{i+1}a_ip_i$.

Let r_i be a point of (p_iq_i) within $\varepsilon_2/3$ of p_ii. It follows from Theorems III.6.D and III.5.C, and a procedure previously used that there is a PL arc $r_ic_ir_{i+1}$ in $\{r_i\} \cup \{r_{i+1}\} \cup$ Int J_i as shown in Figure III.6.A.b and straight arcs joining this

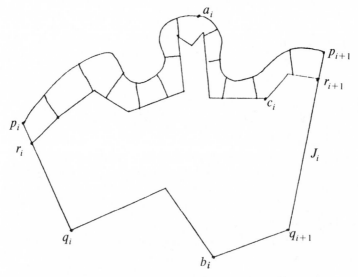

FIGURE III.6.A.b

arc to $p_i a_i p_{i+1}$ as shown so that each of the small upper simple closed curves in Figure III.6.A.b has diameter less than ε_2.

We build such a PL configuration in the PL disk in D bounded by the image of J_i and extend h to take the lower PL disk in Figure III.6.A.b (as well as the straight arcs shown joining $r_i c_i r_{i+1}$ to $p_i a_i p_{i+1}$) into this PL disk. This is done so that the image of each of the small upper simple closed curves in Figure III.6.A.b has a diameter of less than ε_2.

The extension is made for each J_i and we move onto the next step which is a repetition of the last except that we use ε_3 instead of ε_2. The extension is continued and the limit is the required extension. If $p \in \text{Int } J$, h is continuous at p since some neighborhood of p eventually misses the small disks at a far-out stage on which h has not been precisely defined. If $p \in J$, h is continuous at J, since the far-out disks are small and are sent by h (even imprecisely at this stage) into disks bounded by small simple closed curves.   □

With Theorem III.6.A we can lift a restriction from Theorem III.1.1.C.

THEOREM III.6.B. *For each pair of simple closed curves J_1, J_2 in R^2, each homeomorphism ϕ of J_1 onto J_2 can be extended to a homeomorphism h of $J_1 \cup \text{Int } J_1$ onto $J_2 \cup \text{Int } J_2$.*

PROOF. It follows from Theorem III.6.A that there is a homeomorphism of $J_1 \cup \text{Int } J_1$ onto a triangular disk and a homeomorphism g_2 of $J_2 \cup \text{Int } J_2$ onto D that agrees with $g_1 \phi^{-1}$ on J_2. Then $g_2^{-1} g_1 = h$ satisfies the requirements.   □

THEOREM III.6.C. *For each pair of simple closed curves J_1, J_2 in R^2 there is a homeomorphism h of R^2 onto itself such that $h(J_1) = J_2$ and h has compact support.*   □

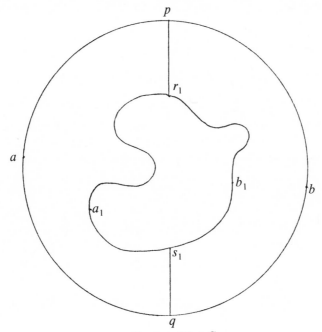

FIGURE III.6.C

THEOREM III.6.D. *Suppose $p_1 p_2$ is an arc of simple closed curve J, U is the bounded component of $R^2 - J$, and $p_1 r_1$, $p_2 r_2$ are straight arcs which lie except for p_1, p_2 in U. Then for each neighborhood O of $p_1 p_2$ there is an arc $s_1 s_2$ in $O \cap U$ such that $p_i r_i \cap s_1 s_2$ for $i = 1, 2$.*

PROOF. Suppose $p_1 r_1 c r_2 p_2$ is a PL arc containing $p_1 r_1 \cup p_2 r_2$ which lies except for its ends in U. Use a brick partitioning (such as shown in Figure III.4.A) of part of R^2 to get a connected open set N such that $r_1 c r_2 \cap N = \emptyset$, $p_1 p_2 \subset N \subset \overline{N} \subset 0$, each component of Bd N is a polygon, while Bd N and $p_1 r_1 c r_2 p_2$ have triangulation that make them in general position. We suppose some point $b \in J$ lies in the unbounded component of $R^2 - \overline{N}$. By cutting slices from N along parts of $p_1 r_1 c r_2 p_2$, we reduce N until Bd $N p_1 r_1 c r_2 p_2 \cap$ Bd N consists of just two points s_1, s_2 of $p_1 r_1 p_2 r_2$, respectively, as shown in Figure III.6.D.

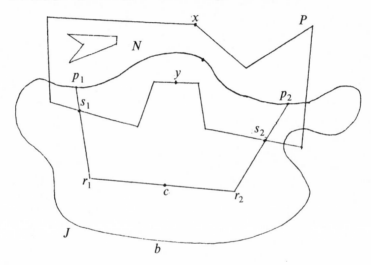

FIGURE III.6.D

The arc $s_1 y s_2$ of Figure III.6.D satisfies Theorem III.6.D so we finish the proof of Theorem III.6.D by showing how to get $s_1 y s_2$. Note that s_1, s_2 belong to the same boundary component P of Bd N since Bd N crosses the simple closed curve $K = p_1 q_1 \cup p j_1 r_1 c r_2 p_2$ only at s_1 and s_2. Suppose s_1 and s_2 separate x from y on P where $y \in$ Int K. Then $s_1 y s_2$ lies except for its ends in Int K. Since Int K does not intersect J, $s_1 y s_2 \subset U$. □

WILD 2-SPHERES

It is natural to try to generalize to higher dimensions interesting concepts and theorems from the plane. One of those that does not generalize is the Schoenflies theorem. There is a 2-sphere in R^3 such that there is no homeomorphism of R^3 onto itself taking the 2-sphere onto the boundary of a round ball. In this chapter we give many such examples.

Perhaps more details are included in this chapter than the reader will want to assimilate before moving to later chapters. Some feel that one should master all previous material before trying to forge ahead but that is not the way that Chapter IV was constructed. The material on the Alexander horned sphere and Antoine's necklace is basic. That on involutions, Fox-Artin spheres, and hooked rugs is for breadth and is not required for work in future chapters. A prudent plan of action might be to become familiar with some of the basic examples and then return to study more complicated ones as one gets better acquainted with the complexities of R^3. Examples of unusual spheres were given by Alexander [A_3, A_4], Alford [A_6, A_7], Ball [B_1], Bing [B_5, B_{13}, B_{31}], Cannon [C_6], Casler [C_7], Daverman and Eaton [DE_1, DE_2], Eaton [E_2], Fort [F_2], Fox and Artin [FA], Hempel [H_7], Martin [M_1, M_2], Rolfsen [R_1], Rosen [R_2].

IV.1. Tame and wild 2-spheres. An n-sphere is an object homeomorphic with the boundary of an $(n + 1)$-ball. The *canonical n-sphere* is the set of points in R^{n+1} satisfying the equation

$$x_1^2 + x_2^2 + \cdots + x_{n+1}^2 = 1.$$

We use S^n to denote this canonical n-sphere. Any object topologically equivalent to this canonical n-sphere is called an n-sphere and on occasions it is convenient to denote one of these other n-spheres by S^n. If B^{n+1} is an $(n + 1)$-*ball* and $=_t$ means is *topologically equivalent to*, then

$$\text{Bd } B^{n+1} =_t S^n.$$

The above definition has more intuitive meaning in lower dimensions. For example, one sees that

$$S^0 =_t \text{pair of points},$$
$$S^1 =_t \text{circle} =_t \text{simple closed curve},$$
$$S^2 =_t \text{boundary of a round ball}.$$

Also, the boundaries of ellipsoids, rectangular solids, and tetrahedra are examples of 2-spheres. The boundary of a rough rock might be a more exotic example. It is more difficult to visualize examples of higher-dimensional spheres. There are not such readily available examples of n-spheres for $n = 3$ in Euclidean 3-space R^3, the space in which we presume we live.

A 2-sphere S^2 in R^3 is called *tame* if there is a homeomorphism of R^3 onto itself that takes S^2 onto the boundary of a round ball. If there is no such homeomorphism, S^2 is called *wild*.

If S^2 is a 2-sphere in R^3, the union of S^2 and its bounded complementary domain is called a *crumpled cube*. (Think of "the cow with the crumpled horn".) If S^2 is wild, there is no assurance that the associated crumpled cube is homeomorphic with a 3-ball. If S^2 lies in S^3, rather than in R^3, the union of S^2 and either of its complementary domains is called a *crumpled cube*. Any set homeomorphic with a crumpled cube is called a crumpled cube irrespective of where it is embedded.

If (X, T) is a complex we may call X a *geometric complex* to emphasize that we are not associating any particular triangulation with it. A subset Y of a geometric complex X is called a *polyhedron* if there is a triangulation T of X (compatible with X's linear structure) such that Y is the union of simplexes of T. In general, a subset W of a geometric complex X is called *tame* if there is a homeomorphism of X onto itself that takes W onto a polyhedron. Recall that we have defined simplexes to be closed and therefore a tame subset of X is closed in X.

IV.2. 3-spheres. A 3-sphere may be more difficult to visualize than a lower-dimensional sphere. The 3-sphere is of primary concern in the study of 3-manifolds. We give special attention to S^3 and describe four models of it.

IV.2.A. S^3 *as the boundary of a 4-simplex.* If one wants to consider S^3 with a PL structure, an approach is to regard S^3 as the boundary of a 4-simplex. Looking at S^3 with the triangulation engendered by the 4-simplex, one would see S^3 as the union of five tetrahedra. A disadvantage of such an approach is that we have used a complicated object (4-simplex) to define a simpler one.

IV.2.B. S^3 *as the union of balls.* It is sometimes convenient to regard S^3 as the union of two 3-balls. Before describing this model, we first consider some lower-dimensional aspects of such a model.

One may decompose a circle into two semicircles. Hence, if one takes two disjoint arcs and uses a homeomorphism to sew the end points of one arc onto the end points of the other, one has an example of a 1-sphere. One can get a model of a 2-sphere by considering two disks D_1, D_2 and joining them along their boundaries. If h is the homeomorphism of Bd D_1 onto Bd D_2 such that each point p of Bd D_1 is fitted to the point $h(p)$ of Bd D_2, we may write

$$S^2 =_t D_1 \cup_h D_2.$$

Similarly, one may obtain a model of S^3 by sewing two balls B_1^3, B_2^3 together with a homeomorphism h of Bd B_1^3 onto Bd B_2^3. The sewing is abstract rather than physical since it cannot be realized in R^3. See Figure IV.2.B. We write

$$S^3 =_t B_1^3 \cup_h B_2^3.$$

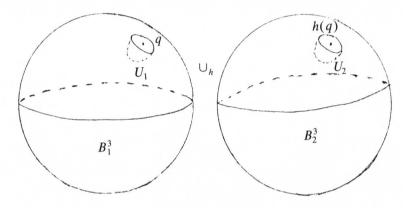

FIGURE IV.2.B

The topology of $B_1^3 \cup_h B_2^3$ is as follows. If $p \in$ Int B_i^3, then any open set in Int B_i^3 containing p is a neighborhood of p in $B_1^3 \cup_h B_2^3$. (We define a *neighborhood* of p as an open set containing p.) If $q \in$ Bd B_1^3, then $\{q\} \cup \{h(q)\}$ is considered a point in $B_1^3 \cup_h B_2^3$. To get a neighborhood of such a $\{q\} \cup \{h(q)\}$, one considers an open set U_1 in B_1^3 containing q and an open set U_2 in B_2^3 containing $h(q)$ such that $h(U_1 \cap$ Bd $B_1^3) = U_2 \cap$ Bd B_2^3. The associated neighborhood of $\{q\} \cup \{h(q)\}$ is

$$U_1 \cup_{h|(U_1 \cap \text{Bd } B_1^3)} U_2.$$

See Figure IV.2.B.

IV.2.C. S^3 *as the one point compactification of* R^3. Before describing S^3 as the one-point compactification of R^3, we consider a 2-dimensional version of this model.

Suppose S^2 is a round 2-sphere tangent to the plane R^2 at a point p. Let q be the point of S^2 diametrically opposite to p. Consider the projection from q of $S^2 - \{q\}$ onto R^2. If $x \in S^2 - \{q\}$, its image is the point where the line joining q and x passes through R^2. Hence

$$S^2 - \{q\} =_t R^2.$$

This suggests that the one point compactification of R^2 is a model for S^2—the point of compactification added to R^2 corresponds to q.

The points of the 1-point compactification of R^3 are the points of R^3 together with an ideal point which we call the point at infinity. Let us use p_∞ to denote this ideal abstract point. The canonical neighborhoods in this enlarged R^3 are of two sorts: (i) the interior of a ball in R^3; (ii) the union of $\{p_\infty\}$ and the complement in R^3 of a closed 3-ball with center at the origin—the larger the 3-ball, the smaller the neighborhood of p_∞.

Actually, it is easy to see that this model of S^3 is equivalent to the one given in the preceding §IV.2.B. Suppose B_1^3 is the ball in R^3 with its center at the origin and radius 1. Then $\{p_\infty\} \cup (R^3 - \text{Int } B_1^3)$ may be considered as a topological ball B_2^3 with center at p_∞. If r is a ray reaching out from the origin and r' is the part of this ray outside Int B_1^3, then $\{p_\infty\} \cup r'$ can be considered a radius of B_2^3. Then the 1-point compactification of R^3 is the union of B_1^3 and B_2^3 sewn together with the identity map of Bd B_1^3 onto Bd B_2^3. See Figure IV.2.C.

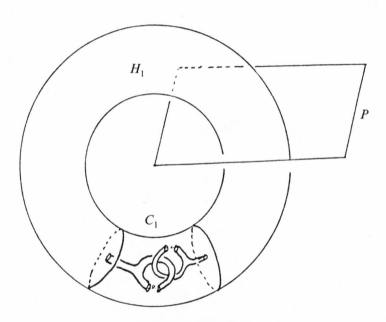

FIGURE IV.2.C

IV.2.D. S^3 *as the union of two solid tori.* Consider two solid tori T_1^3, T_2^3 linked as shown in Figure IV.2.D. Each of these tori is topologically like a doughnut (that is, like $D^2 \times S^1$). Can the two tori be blown up until their boundaries fit? If this could be done, their union would be a model of S^3. There is a bit of a problem in getting one of them to swallow the point at infinity. Perhaps one could visualize this expansion better after considering a more precise abstract approach. We describe one below.

Let M_i be a meridional circle on Bd T_i^3 (M_i bounds a disk cross section D_i of T_i^3 but no disk in Bd T_i^3) and L_i be a longitudinal circle on Bd T_i^3 (L_i bounds a

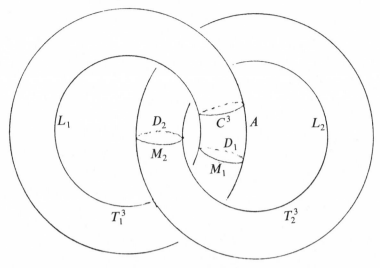

FIGURE IV.2.D

disk in $R^3 - \text{Int } T_i^3$ but not on Bd T_i^3) as shown in Figure IV.2.D. Let h be a homeomorphism of Bd T_1^3 onto Bd T_2^3 that takes M_1 onto L_2 and L_1 onto M_2. Then

$$S^3 =_t T_1^3 \cup_h T_2^3.$$

The topology of $T_1^3 \cup_h T_2^3$ is as described in §IV.2.B.

This model of S^3 is a disguised version of the union of two balls given in §IV.2.B. If one thickens the disk D_1 to get a pillbox (solid cylinder) C^3 whose lateral surface is an annulus A going meridionally around Bd T_1^3, the closure of $T_1^3 - C^3$ is a 3-ball B_1^3, i.e.,

$$B_1^3 = \overline{T_1^3 - C^3}.$$

Also,

$$B_2^3 = C^3 \cup_{h|A} T_2^3$$

is a 3-ball and

$$S^3 =_t T_1^3 \cup_h T_2^3 =_t B_1^3 \cup_{h'} B_2^3$$

where

$$h' = \begin{cases} h \text{ on Bd } T_1^3 - A, \\ \text{Identity on bases of } C^3. \end{cases}$$

Other models of S^3 are given by Bing in $[\mathbf{B_{26}}]$.

IV.2.E. S^3 *is a special sort of 3-manifold.* An *n-manifold* is a separable metric space each point of which has a neighborhood homeomorphic to Euclidean *n*-space R^n. An *n-manifold-with-boundary* is a separable metric space each of whose points has a neighborhood whose closure is topologically equivalent to an

n-simplex. Hence, an n-manifold is an n-manifold-with-boundary but not con-versely. Some authors call a compact n-manifold (without boundary) a *closed manifold*.

Note that S^3 is compact connected 3-manifold. However, there are many other compact connected 3-manifolds topologically different from S^3. It is easy to ask what additional restrictions placed on a 3-manifold makes it a 3-sphere. The answers may be elusive.

Bing has shown [$\mathbf{B_{10}}$] that a compact connected 3-manifold is a 3-sphere if each simple closed curve in it lies in a topological 3-ball in it. It is known [$\mathbf{B_4}$] to be S^3 if it has a *sequentially unicoherent triangulation* T—the 3-simplexes in T can be ordered $\Delta_1^3, \Delta_2^3, \ldots, \Delta_n^3$ such that for $2 \leqslant j \leqslant n$, $\Delta_j^3 \cap \left(\cup_{i<j} \Delta_i^3 \right)$ is connected. These results are discussed in Chapter XVIII. The Poincaré conjecture is that a compact connected 3-manifold is S^3 if it is 1-connected as defined in §IV.4, but the result is in doubt. The paper [$\mathbf{W_4}$] contains a retraction of a mistake [$\mathbf{W_3}$]. It was once claimed [$\mathbf{P_4}$] that a compact connected triangulated 3-manifold was topologically S^3 if it had the property that 1-cycles in it (as discussed in Chapter VII) bounded in it, but that error was soon rectified [$\mathbf{P_5}$].

IV.3. Alexander horned sphere. In the early part of this century [$\mathbf{A_1}$] it was erroneously claimed that if S^2 is a 2-sphere in R^3, then there is a homeomorphism of R^3 onto itself that takes S^2 onto the boundary of a round ball. This would have implied that each 2-sphere in R^3 is tame and that the Schoenflies theorem generalizes to R^3. It was perhaps in trying to tighten a "proof" of this claim that Alexander discovered the Alexander horned sphere (partially shown in Figure IV.3) and attacked a PL version of the Schoenflies theorem for R^3. See [$\mathbf{A_3, A_2}$].

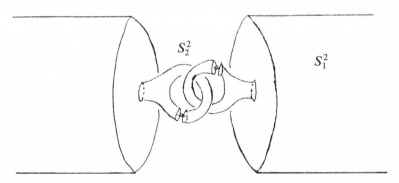

FIGURE IV.3

The Alexander horned sphere shown in Figure IV.3 can be described as the limit of a sequence of tame 2-spheres $S_1^2, S_2^2, \ldots$. The first of these tame 2-spheres S_1^2 is obtained by taking a long 2-dimensional cylinder closed at both ends and bending it until the ends are near each other and parallel. Figure IV.3 shows the ends of the cylinder but not the central part. The second approximation S_2^2 is obtained by pushing tubes out of each end until they almost hook as shown. Further approximations $S_3^2, S_4^2, \ldots$, are obtained by pushing out additional tubes, pushing out additional tubes, etc. The resulting limit has the property that although it is topologically equivalent to the boundary of a round ball, the unbounded complementary domain of it is not topologically equivalent to the exterior of a round 3-ball.

EXERCISE IV.3.A. Describe a homeomorphism between an Alexander horned sphere of Figure IV.3 and a round 2-sphere.

Solution. Let D_0, D_1 be two small mutually disjoint disks on a round 2-sphere S^2; $D_{i,j}$ $(i = 0, 1; \ j = 0, 1)$ be four very small mutually disjoint disks with $D_{ij} \subset \text{Int } D_i$; D_{ijk} eight very very small disks;.... . A homeomorphism h of the Alexander horned sphere A described above onto S^2 sends $A \cap S_1^2$ onto $S^2 -$ $\cup$ Int D_i, $A \cap S_2^2$ onto $S^2 - \cup$ Int D_{ij}, $A \cap S_3^2$ onto $S^2 - \cup$ Int D_{ijk},.... . If $p \in A - \cup S_i^2$, let U_t be the component of $A - S_t^2$ containing p and D_t be the $D_{ij\cdots k}$ bounded by $h(\text{Bd } U_t)$. Then $h(p) = \cap D_t$. $\square$

EXERCISE IV.3.B. The set of points of the Alexander horned sphere not in any of the approximating S_i^2's is called "*the tips of the horns*". To what is this set topologically equivalent?

Answer. A Cantor set. $\square$

The union of the Alexander horned sphere and its bounded complementary domain is a crumpled cube. We could have modified our description of the Alexander horned sphere A as the limit of S_i^2's to get the crumpled cube bounded by A as the intersection of handlebodies H_1, H_2,.... .

EXERCISE IV.3.C. Express $A \cup \text{Int } A$ as the intersection of handlebodies.

Answer. Let H_1 be the solid torus in Figure IV.3.C. It is topologically equivalent to a doughnut or the cartesian product $B^2 \times S^1$. Let C_1 be a thin meridional slice of H_1 as shown in Figure IV.3.C. The handlebody H_2 is obtained from H_1 by removing C_1 and adding back the ends of C_1 together with two hooked eye bolts in C_1 as shown where one of the eye bolts fits onto one end of C_1 and the other the opposite ends. An *eye bolt* is a solid torus on a stem.

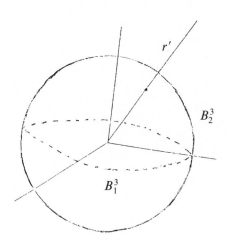

FIGURE IV.3.C.

To get H_3, one removes a meridional slice from each eye bolt and makes a replacement as before. The procedure is continued. Then $A \cup \text{Int } A = \cap H_i$. □

Crumpled cubes bounded by horned spheres come in many shapes and sizes. Figure IV.3.D shows the emblem on the front of many T-shirts. Figure IV.3.E shows a version suggested by Bill Eaton. Each of the examples described by Figures IV.3.D, IV.3.E has the property that there is a homeomorphism h: $R^3 \to R^3$ that takes Alexander's original example onto the examples. Each suggests that the Schoenflies theorem about 1-spheres in R^2 does not extend to 2-spheres in R^3.

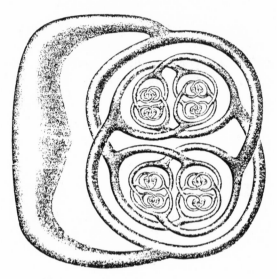

Reprinted with permission of the publisher from Bing, R., *Some aspects of the topology of 3-manifolds related to the Poincare conjecture*, Lectures on Modern Mathematics, vol. 2, edited by T. L. Saaty, John Wiley & Sons, Inc., New York, 1964, p. 117.

FIGURE IV.3.D

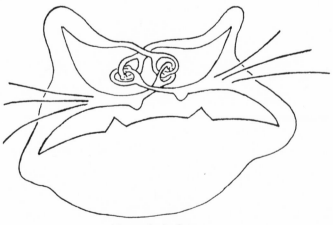

Alexander's Cat

FIGURE IV.3.E

IV.4. Simple connectivity. We examine some properties that can be used to show that Alexander's horned sphere is wild.

A set X is 1-*connected* if each map of Bd Δ^2 into X can be extended to a map of Δ^2 into X. A set which is 1-connected is also called *simply-connected*. If $f: Y \to X$, we say that f is *homotopic to a constant* if there is a map $g: Y \times [0, 1] \to X$ such that $g(y \times 0) = f(y)$ and $g(Y \times 1)$ is a point. We say that a set $Y \subset X$ *can be shrunk to a point in X* if the identity map of Y into X is homotopic to a constant. We regard t as a parameter and Y as having been shrunk to $g(Y \times t)$ as time t. A necessary and sufficient condition that a set Y can be shrunk to a point in a space X is that Y is the base of a singular cone in X.

THEOREM IV.4.A. *The complement of the crumpled cube bounded by the Alexander horned sphere is not simply-connected.*

This theorem follows from Theorem VI.10.C from Chapter VI.   □

COROLLARY IV.4.B. *The Alexander horned sphere is wild.*   □

EXERCISE IV.4.C. Give a homeomorphism of a round 3-ball onto the union of the Alexander horned sphere and its bounded complementary domain.

Solution. Let X_1, X_2 be two end slices of a 3-ball B^3 as shown in Figure IV.4.B. A description of a suitable homeomorphism h can be started by saying that h sends the part of B^3 between X_1 and X_2 to $H_1 - C_1$ of Figure IV.3.B. Let X_{i1}, X_{i2} be small mutually disjoint end slices of B^3 in X_i ($i = 1, 2$) as shown. Extend h to take $B^3 - \cup X_{ij}$ to H_2 minus two meridional cross sections of handles of H_2. If this procedure is continued one gets a homeomorphism h from most of B^3 to most of the crumpled cube. It is not yet defined on the Cantor set $(\cup X_i) \cap (\cup X_{ij})$ $\cap (X_{ijk}) \cap \cdots$ but this set is sent by h onto the tips of the horns. Had we used $C_{1ij \cdots k}$'s to designate the thin meridional slices in the H's, one could have defined the image of this Cantor set to be $C_1 \cap (\cup C_{1i}) \cap (\cup C_{1ij}) \cap \cdots$.   □

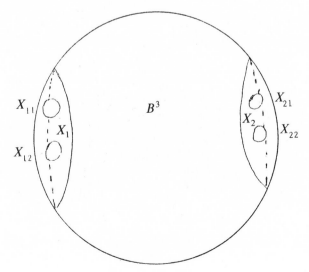

FIGURE IV.4.B

The above exercise shows it is not the interior of this crumpled cube that topologically distinguishes the cube from a round 3-ball. However, the exterior of this crumpled cube is not simply-connected. The exterior of a tame cube is simply-connected.

EXERCISE IV.4.D. If B^3 is a round 3-ball in R^3 and P is a polygon in $R^3 - B^3$, describe a map f that shows that P can be shrunk to a point in $R^3 - B^3$.

Solution. Let g be a PL map of Δ^2 into R^3 such that g takes Bd Δ^2 homeomorphically onto P. Let $p_0 \in B^3 - g(\Delta^2)$ and B^+ be a round 3-ball such that $B^3 \subset$ Int $B^+ \subset R^3 - P$. Let ϕ be the projection from p_0 of $B^+ - p_0$ onto Bd B^+. A suitable map $f: \Delta^2 \to R^3 - B$ can be described as follows: $f(x) = g(x)$ if $g(x) \in R^3 - B^+$; $f(x) = \phi g(x)$ if $g(x) \in B^+$. $\square$

IV.5. Solid Alexander horned sphere. If an Alexander horned sphere A is placed in S^3 rather than in R^3 and U is the nonsimply-connected complementary domain of A, any set homeomophic to $A \cup U$ is called a *solid Alexander horned sphere*. It is a crumpled cube that is topologically different from a 3-ball.

We could have defined a solid Alexander horned sphere in R^3 rather than in S^3 if we had defined the horned sphere inside out with the tubes reaching in instead of out. We give such a description that has been useful.

Let C be a two-dimensional right circular cylinder in R^3 with bases D_1, D_2. See Figure IV.5. Two mutually exclusive disks in each D_i ($i = 1, 2$) are replaced by the surfaces of tubes T_{i1}, T_{i2}, and disks D_{i1}, D_{i2} as shown in Figure IV.5 where D_{i1}, D_{i2} are the bases of a right circular cylinder C_i and $D_1 \cup T_{11} \cup C_1 \cup T_{12}$ is hooked to $D_2 \cup T_{21} \cup C_2 \cup T_{22}$ as shown.

Disks in the bases of the cylinder C_i ($i = 1, 2$) are replaced by the surfaces of tubes T_{i11}, T_{i12}, T_{i21}, T_{i22} and disks D_{i11}, D_{i12}, D_{i21}, D_{i22} as before. The process is continued to get the horned sphere M. We use M_1 to denote the part of M that is the closure of the part of M on the exterior of $C_1 \cup C_2$. Then $M_1 = C -$ (union

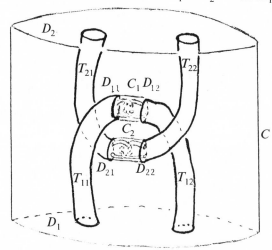

FIGURE IV.5

of four disks, two from each of D_1, D_2) $\cup$ (union of four tubes, T_{ij} ($i = 1, 2$; $j = 1, 2$)). Likewise, M_i denotes the closure of the part of M on the exterior of $\cup C_{j_1 j_2 \cdots j_i}$. It is topologically equivalent to a 2-sphere minus 2^{i+1} open disks. Let M_0 be the Cantor set $M - \cup M_i$. It is the tips of the horns of M.

Although M is homeomorphic to a round 2-sphere, its bounded complementary domain is not simply-connected. We call the complementary domain of M that is not simply-connected the bad complementary domain of M. Note that the union of M and its bad complementary domain U is a solid Alexander horned sphere.

IV.6. Peculiar involutions. We digress from our treatment of 3-manifolds to list some interesting properties of crumpled cubes that will not be considered in depth in this book. A result by Bing [**B$_5$**] that we shall not prove but which has had applications is the following.

THEOREM IV.6.A. *The union of two solid Alexander horned spheres sewed together with an identity-like homeomorphism on their boundaries is topologically S^3 in the sense that there are two topological solid Alexander horned spheres K_1, K_2 in S^3 and a homeomorphism h: $K_1 \rightarrow K_2$ such that*

$$S^3 = K_1 \cup K_2,$$
$$\text{Bd } K_1 = \text{Bd } K_2 = K_1 \cap K_2, \text{ and}$$
$$h \mid \text{Bd } K_1 = identity.$$

It would be interesting to be able to visualize a picture of Bd $K_1 = $ Bd $K_2 = K_1 \cap K_2$ of Theorem IV.6.A. The tips of the horns from one side would look like the tips of the horns from the other side.

Smith has shown [**S$_4$**] that for any orientation preserving periodic homeomorphism of S^3 onto itself, the fixed point set is either null, a 1-sphere, or S^3 itself; while for any orientation reversing periodic homeomorphism, the fixed point set is either a 0-sphere or a 2-sphere. He also has shown that if h is a periodic homeomorphism of S^3 whose fixed point set is a 2-sphere, then h is an *involution* (homeomorphism of period 2). Theorem IV.6.A was the key link in showing that there is an involution of S^3 onto itself whose fixed point set is a wild 2-sphere.

THEOREM IV.6.B. *There is an involution S^3 whose fixed point set is a wild 2-sphere.* □

A homeomorphism of R^n onto itself is a *rigid motion* if it is an isometry. Examples of rigid motions of R^3 are translations, rotations about lines, reflections through planes, lines, or points, and compositions of such motions. A homeomorphism h is *equivalent to a rigid motion* if there is a homeomorphism f and a rigid motion r such that $h = f^{-1}rf$. In general, two homeomorphisms h_1, h_2 are *equivalent* if there is a homeomorphism f such that $h_1 = f^{-1}h_2 f$.

THEOREM IV.6.C. *There is an orientation reversing involution of R^3 which is inequivalent to any rigid motion.* □

Once it had been shown that one could obtain S^3 by sewing two solid Alexander horned spheres together with an identity-like homeomorphism on their boundaries it was found that there were many other crumpled cubes—each topologically different from the others—such that if one took any of them and sewed it onto a topological copy of itself with an identity-like homeomorphism on their boundaries, then S^3 was obtained. See Alford [A_7], Bing [B_{20}], Montgomery and Samehon [MS], and Montgomery and Zippin [MZ]. Other sewings of crumpled cubes are given by Ball [B_1], Bing [B_8], Cannon [C_6], Casler [C_7], Daverman and Eaton [DE_1, DE_2], Eaton [E_2], and Martin [M_1].

THEOREM IV.6.D. *There is an uncountable family of mutually inequivalent orientation reversing involutions of R^3 onto itself.* $\square$

IV.7. Antoine's necklace. Consider in R^3 a decreasing sequence of sets T_1, $\cup T_{1i}$, $\cup T_{1ij}$,..., such that T_1 is a topological solid torus; $\cup T_{1i}$ is the union of a collection of topological solid tori $\{T_{1i}\}$ such that these solid tori form a chain in Int T_1 that go around T_1 as shown in Figure IV.7.A; $\cup T_{1ij}$ is the union of a collection of solid tori $\{T_{1ij}\}$ such that for each T_{1i_0}, the T_{1i_0j}'s form a chain which go around T_{1i_0}; $\cup T_{1ijk}$ is the union of topological tori T_{1ijk};.... The limit as n increases without limit of the mesh of $\{T_{1ij\cdots n}\}$ is 0. The *mesh* of a collection of sets in a metric space (like the holes in a screen wire) is the least upper bound of the diameters of the objects. The intersection of the sets $\cup T_1$, $\cup T_{1i}$,..., is called an Antoine's necklace.

Although Figure IV.7.A shows four T_{1i}'s in T_1, we do not make this restriction. For all we care, the number of links in the chains may vary except that in each case it exceeds two.

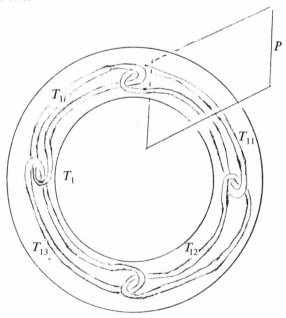

FIGURE IV.7.A

EXERCISE IV.7.A. Show that irrespective of how the number of links of the chains vary, there is a homeomorphism h of the resulting Antoine's necklace onto a Cantor set.

Solution. In building a Cantor set on an interval one is not required to remove an open middle third at each stage and be left with two pieces, but instead could remove several open holes and be left with several small pieces. Divide an interval into $2n_1 - 1$ equal pieces where n_1 is the number of tori in $\{T_{1i}\}$. Delete the interiors of the even pieces and agree that h sends T_{1i} into the ith remaining piece. As a next step divide the ith remaining piece into $2n_{i2} - 1$ pieces where n_{i2} is the number of tori of $\{T_{1ij}\}$ in T_{1i}; remove the interiors of the even ones and let h send T_{1ij} into the jth of the remaining. The description of h is continued. Although no step reveals precisely where a point goes, the limit does. □

Antoine [A_8] described the necklace to illustrate the fact that even compact 0-dimensional sets in R^3 which are topologically alike might have topologically different complements. He did not actually prove that the complement of Antoine's necklace was not simply-connected but he assumed this. It was shown by Coelho [C_9] that the complements of certain Antoine's necklaces were not simply connected.

Someone has asked why the set is called an Antoine's necklace. Perhaps the answer is that it was first described by Antoine [A_8]. Like a necklace its components are small and hold together. If it were slung over one's arm, not a point would drop to the floor. Like a string of beads, it is a thing of beauty with mostly aesthetic value. It has been useful to the mathematician as an example to illustrate certain topological properties.

It may be convenient to think of the $T_{1ij\cdots n}$'s as tubular neighborhoods with small cross radius of smooth simple closed curves, but we only required that they be topologically equivalent to such objects and not that they be thin. We did require that the mesh of the $\{T_{1ij\cdots n}\}$'s go to zero as the length of $1ij\cdots n$ increased without limit.

We examine the complement of Antoine's necklace by considering the intersections of PL surfaces. Recall from §II.3 that two complexes (X_1, T_1), (X_2, T_2) in R^3 are in *general position* (GP) if not vertex of T_i $(i = 1, 2)$ lies on a vertex, 1-simplex or 2-simplex of T_j $(i \neq j)$ and no 1-simplex of T_1 intersects any 1-simplex of T_2. If (X_1, T_1), (X_2, T_2) are not GP, slight pushes at vertices of T_2 give a linear homeomorphism of $(X_2, T_2) \rightarrow (X_3, T_3)$ such that (X_1, T_1) and (X_3, T_3) are GP. If X_1, X_2 are geometric complexes rather than complexes we say X_1, X_2 are GP if X_1, X_2 have triangulations T_1, T_2 such that (X_1, T_1), (X_2, T_2) are GP. Other versions of general position are found in §IX.1.

Before showing that Antoine's necklace is wild we borrow some concepts related to Chapter VI. Figure IV.2.D shows two linked solid tori T_1^3, T_2^3. For each $i = 1, 2$, M_i is a meridional simple closed curve on Bd T_i^3 and L_i is a longitudinal simple closed surve on Bd T_i^3.

If f is the map of a simple closed curve J into Bd T_i^3, we say that $f(J)$ *circles* Bd T_i^3 *meridionally* (or *longitudinally*) if it crosses L_i (or M_i) algebraically a nonzero number of times. The following result follows from Theorem II.3.C and the techniques of Chapter VI.

THEOREM IV.7.B. *Suppose* T_1^3, M_1, T_2^3, L_2 *of Figure* IV.2.D *are PL and f is a PL map of* Δ^2 *into* R^3 *such that* $f(\text{Bd } \Delta^2) \subset \text{Bd } T_1^3$ *and* $f(\Delta^2)$ *is GP with respect to* Bd T_2^3. *If* $f(\text{Bd } \Delta^2)$ *circles* Bd T_1^3 *longitudinally, then* Δ^2 *contains a PL disk* D^2 *such that* $f(\text{Bd } D^2)$ *circles* T_2^3 *meridionally.* □

THEOREM IV.7.C. *If f is a map of a disk* Δ^2 *into* R^3 *that sends* Bd Δ^2 *homeomorphically onto* P *of Figure* IV.7.A, *then* $f(\Delta^2)$ *intersects the Antoine's necklace suggested by Figure* IV.7.A.

PROOF. Suppose contrariwise that there is a map f of Δ^2. The $f(\Delta^2)$ intersects only a finite number of the Bd $T_{1ij\cdots n}$'s. We suppose with no loss of generality that each Bd $T_{1ij\cdots n}$ is PL, f is PL, and $f(\Delta^2)$ is GP with respect to each Bd $T_{1ij\cdots n}$.

It follows from Theorem II.3.C and techniques in Chapter VI that Δ^2 contains a subdisk D_1 such that $f(D_1)$ circles Bd T_1 meridionally. It follows from Theorem VI.10.I that D_1 contains a disk D_2 such that $f(\text{Bd } D_2)$ circles one of Bd T_{11}, Bd T_{12}, Bd T_{13}, Bd T_{14} either meridionally or longitudinally. However, if $f(\text{Bd } D_2)$ circles Bd T_{1i} longitudinally it follows from Theorem IV.7.B that D_2 contains a subdisk D_3 such that $f(D_3)$ circles Bd $T_{1(i\pm1)}$ meridionally.

Inductively we find that at each stage $1ij\cdots n$ there is a disk D in Δ^2 and $T_{1ij\cdots n}$ such that $f(D)$ circles Bd $T_{1ij\cdots n}$ meridionally. Hence, $f(D)$ intersects Antoine's necklace suggested by Figure IV.7.A. □

The arguments for Theorem IV.7.C applied equally well to Antoine necklaces that had three or more $T_{1ij\cdots n}$'s at each stage in the T at the preceding stage as suggested by Figure IV.7.A. We note another property that distinguishes these Antoine necklaces from tame Cantor sets.

A set X is *n-locally connected* (*n*-LC) at a point p if for each neighborhood U of p there is a neighborhood V of p such that each map of Bd Δ^{n+1} into $X \cap V$ can be extended to a map of Δ^{n+1} into $X \cap U$. A set is called *locally simply-connected* if it is 1-LC and *locally arcwise connected* if it is 0-LC.

THEOREM IV.7.D. *The complement of Antoine's necklace* A *suggested by Figure* IV.7.A *is not 1-LC at all points of Antoine's necklace.*

PROOF. Let D be a PL disk bounded by P of Figure IV.7.A. Assume $R^3 - A$ is 1-LC at each point of A. Let $\{V_\alpha\}$ be a collection of open sets covering $D \cap A$ such that each map of Bd Δ^2 into a $(R^3 - A) \cap V_\alpha$ can be extended to map Δ^2 into $R^3 - (A \cup P)$. Cover $A \cap D$ with a finite collection of mutually disjoint disks $D_1, D_2, \ldots, D_n$ in Int D such that each D_i lies in a V_α. Get a singular disk D' by replacing each D_i with the image of a disk into $R^3 - (A \cup P)$. Apply Dehn's lemma (see Chapter XVI) to the singular disk D' to get a disk in $R^3 - A$ bounded by P. The assumption that $R^3 - A$ is 1-LC at each point of A leads to a contradiction of Theorem IV.7.C. □

EXERCISE IV.7.E. Suppose C is a Cantor set on a line L in R^3 and D is a disk in R^3 whose boundary misses C. Describe a map f that shows that Bd D bounds a disk in $R^3 - C$.

Solution. Let $B_1, B_2, \ldots, B_n$ be a finite collection of mutually disjoint round 3-balls in $R^3 -$ Bd D such that $C \subset \cup$ Int B_i and $p_i \in B_i - D$. Let ρ_i be a homeomorphism of B_i onto itself such that ρ_i is fixed on Bd B_i and pushes $B_i \cap D$ so close to Bd B_i that $C \cap \rho_i(B_i \cap D) = \varnothing$. Then f can be defined as follows: $f(x) = x$ if $x \in D - \cup B_i$; $f(x) = \rho_i(x)$ if $x \cap B_i$. We project $D \cap (B_i)$ to near Bd B_i. $\square$

IV.8. An Antoine wild sphere. Alexander $[A_3]$ used an Antoine's necklace to build a wild sphere. Suppose that as shown in Figure IV.8, one has an Antoine necklace on the left and a 2-sphere on the right. A disk is removed from the 2-sphere and replaced by a thin *feeler* (capped open cylinder) which runs over to Bd T_1. Then disks are removed from the part of the feeler on Bd T_1 and feelers are run inside T_1 over to each Bd T_{1i}. The process is continued. The limit is called an *Antoine wild sphere*.

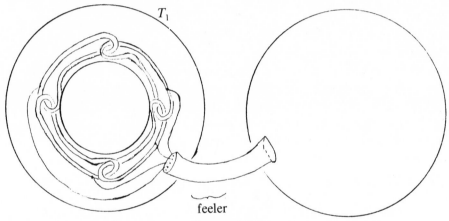

T_1

feeler

FIGURE IV.8

Perhaps Alexander's horned sphere is better known than Antoine's. Why? Perhaps it is that Antoine's is described with verbage $[A_4]$ while Alexander's was given with a meaningful diagram $[A_3]$. Aspects of the following theorem are considered again in Chapter VI.

THEOREM IV.8. *The Antoine wild sphere is wild.* $\square$

IV.9. Some Fox-Artin spheres. Fox and Artin described some interesting wild spheres in [FA]. These spheres were the boundaries of 3-balls formed by thickening the interiors of certain wild arcs. We shall consider three of these wild spheres that have been widely used as examples.

IV.9.A. *A 2-sphere with nonsimply connected complement.* The dotted portion of Figure IV.9.A represents two solid cones with a common base $F_{-1} \cap F_1$ subdivided into a countable collection of frustrums $F_1, F_{-1}, F_2, F_{-2}, F_3, \ldots$. A more

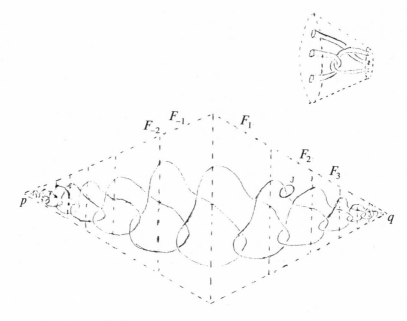

FIGURE IV.9.A

detailed frustrum F_i is shown in the right part of the figure. These frustrums are not a part of the 2-sphere but only serve as guides. The 2-sphere intersects each of these frustrums in the union of three tubular cylinders in a fashion as shown on the right of the figure. In the main part of Figure IV.9.A, we only show the center of these cylinders. If the interior of the oscillating arc pq from p to q is fattened, one obtains the 3-ball whose boundary is a wild sphere which is locally smooth except at two points.

Using techniques of knot theory such as treated in Chapter VI it can be shown that the simple closed curve J shown in frustrum F_2 of Figure IV.9.A cannot be shrunk to a point in $R^3 - pq$. Hence the 3-ball obtained by fattening pq at its interior points is wild as is its 2-sphere boundary.

IV.9.B. *A wild sphere whose complement is topologically like that of a 2-sphere.* Consider the 2-sphere shown in Figure IV.9.B. It is the union of a disk in $F_{-2} \cap F_{-1}$ and a part of the 2-sphere of §IV.9.A. Except at the left side of the sphere, we showed the center lines of the tubular cylinders rather than the cylinders themselves. We use B to denote the 3-ball bounded by the 2-sphere.

Showing that a set X is not 1-LC at a point p is a bit like playing a game with a devil. You pick a neighborhood U of p and defy the devil to pick a neighborhood V of p. Then show that no matter what V the devil picks, you can find a map f: Bd $\Delta^2 \to X \cap V$ that cannot be extended to take Δ^2 into $X \cap U$.

EXERCISE IV.9.B.a. Assume that J of Figure IV.9.A cannot be shrunk to a point in $R^3 - pq$ and that you are playing a game with a devil to show $R^3 - B$ is not 1-LC at q. How could you select U? After the devil has picked V, how could you select f: Bd $\Delta^2 \to (R^3 - B) \cap V$?

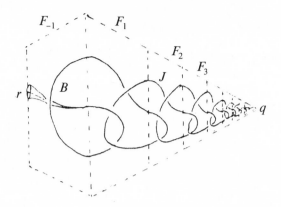

FIGURE IV.9.B

Solution. One selection (there are others) for U is the half space to the right of the plane through $F_1 \cap F_2$. After the devil has picked V pick an integer so large that $F_n \subset V$ and let P be a polygon that links B in F_n as J linked pq in F_2. If P could be shrunk to a point in $(R^3 - B) \cap U_1$, it could be shrunk to a point in $R^3 - pq$. However, it cannot be shrunk in $R^3 - pq$ because J cannot be shrunk and J can be slid (isotoped) to P in $R^3 - pq$. $\square$

EXERCISE IV.9.B.b. Suppose q, B are as shown in Figure IV.9.B and Bd B is locally PL except at q. Suppose $g: \Delta^2 \to R^3 - \{q\}$ is such that $B \cap g(\text{Bd } \Delta^2) = \varnothing$. Explain how to get a map $f: \Delta^2 \to R^3 - B$ which agrees with g on Bd Δ^2.

Solution. Enlarge B in $R^3 - \{q\}$ to get a 3-ball B^+ such that $B \subset \{q\} \cup \text{Int } B^+ \subset B^+ \subset R^3 - g(\text{Bd } \Delta^2)$. Let D be a disk in Bd $B^+ - \{q\}$ that contains (Bd B^+) $\cap g(\Delta^2)$ and ϕ be a map of Δ^2 into D that agrees with g on $g^{-1}(D)$. One could use the Tietze extension theorem to get such a ϕ. Then $f(x) = g(x)$ or $\phi(x)$ according as x lies or does not lie in the same component of $\Delta^2 - g^{-1}(D)$ as Bd Δ^2. $\square$

The 3-ball B shown in Figure IV.9.B has the peculiar property that although it is wild, its complement is homeomorphic to R^3 minus the origin. A subset X of R^3 is *cellular* if it is the intersection of a decreasing sequence of 3-balls in R^3 such that each lies in the interior of the preceding.

THEOREM IV.9.B.c. *B is cellular*.

INDICATION OF PROOF. Suppose rq is a proper subarc of pq of §IV.9.A and U is an open set that contains rq. To show that B is cellular it suffices to show that there is a 3-ball in U containing rq.

To show the existence of such a 3-ball containing rq we first consider a small 3-ball B_1 in U about q so that Int B_1 contains all of rq except an arc rb and an arc ac as shown in the top part of Figure IV.9.B.c. The figure does not show all the wiggles in rb since they are not relevant.

The 3-cell B_1 is enlarged in U to a 3-ball B_2 shown in the second part of Figure IV.9.B.c so that all of rq except a short subarc $a'c$ is covered as shown. Two feelers are extended from B_1 to form B_2. Next, a tunnel is bored into B_2 so that the arc $a'cbr$ is not covered by the resulting 3-ball B_3. Then a feeler is pushed out

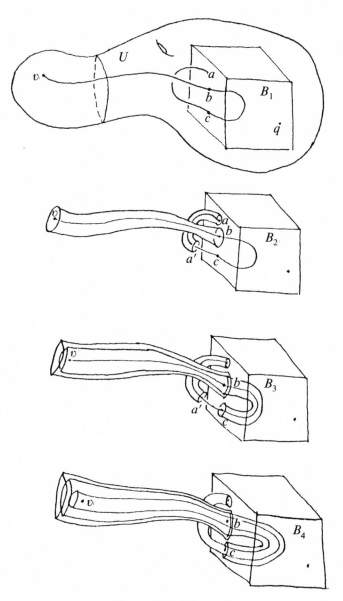

FIGURE IV.9.B.c

from a' so that all of rq gets covered as shown in the lower part of Figure IV.9.B.c. It may be noted that a straight line drawn to point r from the left passes through Bd B_3 at least three times.   □

IV.9.C. *A wild 2-sphere whose complement is simply-connected but topologically different from the complement of a tame 2-sphere.* See Figure IV.9.C. Again we show the center lines of the tubular cylinders rather than the cylinders. Let P be the plane containing $F_{-1} \cap F_1$, X be the part of pq of §IV.9.A on or to the right of P and X' be the reflection of X through P. Then $X \cup X'$ is the union of an arc

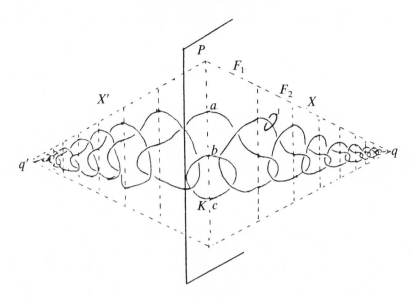

FIGURE IV.9.C

$q'aq$ and a simple closed curve K. The wild 2-sphere of this section is the boundary of the 3-ball B obtained by fattening $q'aq$ along its interior. We shall indicate why $R^3 - B$ is simply-connected and $R^3 - B$ is topologically different from R^3-point.

EXERCISE IV.9.C.a. Explain how to shrink J of Figure IV.9.C to a point in $R^3 - q'aq$.

Solution. Slide J along the tube about $q'aq$ until it comes into plane P near a. Then enlarge it in P so that it can be pulled over either end. □

One property that makes B wild is that $R^3 - B$ is not 1-LC at q' or q. The following two theorems lead us to another property.

EXERCISE IV.9.C.b. Show that if J of Figure IV.9.C can be shrunk to a point in $R^3 - (K \cup q'aq)$ then J of Figure IV.9.A can be shrunk to a point in $R^3 - pq$.

Solution. Let ϕ be the map of R^3 onto itself that is the identity to the right of the plane P of Figure IV.9.C and the reflection through P elsewhere. If f: $\Delta^2 \to R^3 - (K \cup q'aq)$ shows that J can be shrunk to a point in $R^3 - (K \cup q'aq)$ then ϕf shows that J of Figure IV.9.A can be shrunk in $R^3 - pq$. □

What we really show by this exercise is that J of Figure IV.9.C cannot be shrunk to a point in $R^3 - (K \cup q'aq)$ because we have already mentioned that J of Figure IV.9.A cannot be shrunk to a point in $R^3 - pq$.

EXERCISE IV.9.C.c. Explain why no PL ball in $R^3 - q'aq'$ contains K.

Solution. Suppose C is a 3-ball whose interior contains K and J is so small that $J \cap C = \varnothing$. It follows from Exercise IV.9.C.a that J can be shrunk to a point in $R^3 - q'aq$ with a map $f: \Delta^2 \to R^3 - q'aq$. Let ϕ be a map of Δ^2 into Bd C that agrees with f on $f^{-1}(\text{Bd } C)$. Let g be the map of Δ^2 into $R^3 - (\text{Int } C \cup q'aq)$ which is $f(x)$ or $\phi(x)$ according to whether x lies in or does not lie in some component of $\Delta^2 - f^{-1}(\text{Bd } C)$ containing Bd Δ^2. Then g shrinks J in $R^3 - (K \cup q'aq)$. This violates Exercise IV.9.C.a and the fact that J of Figure IV.9.A cannot be shrunk to a point in $R^3 - pq$. $\square$

The *lamp cord trick* is the observation that if M^3 is a hollow ball (homeomorphic image of $S^2 \times [0,1]$) and for $i = 1$ or 2, T_i is a tube joining the boundary components of Bd M^3 then $\underline{M^3 - T_1} =_t \underline{M^3 - T_2}$ irrespective of whether or not T_2 is "knotted". The set $\underline{M^3 - T_i}$ resembles the closure of a room minus a light bulb on a cord as shown in Figure IV.9.C.d and the homeomorphism yielding the light cord trick comes from untying the knotted cord.

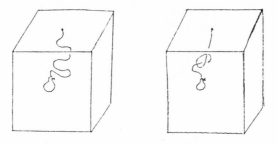

FIGURE IV.9.C.d

It follows from the lamp-cord trick that if X is a compact closed subset of R^3 and B^3 is a round 3-ball in the unbounded component of $R^3 - X$, then X lies in a tame 3-ball in $R^3 - B^3$. Figure IV.9.C.e shows C^3 as a large round ball whose interior contains $B^3 \cup X$. Since B^3 is in the unbounded component of X, there is a "lamp cord" from Bd B^3 to Bd C^3 in $C^3 - X$.

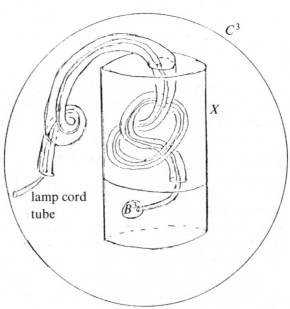

FIGURE IV.9.C.e

IV.10. Bing's hooked rug. Each of the 2-spheres we have previously considered in this chapter contained a wild arc. We now describe a 2-sphere whose wildness is so spread about that each arc in it is tame. The sphere is called *Bing's hooked rug* and is described in [**B**$_{13}$].

The hooked rug S_∞ is simultaneously the limit of a sequence of 2-spheres $S_1, S_2, \ldots$, and the intersection of a decreasing sequence of 3-manifolds-with-boundary $W_1, W_2, \ldots$. We describe a sequence $S_1, W_1, S_2, W_2, \ldots$.

Let S_1 be a tame 2-sphere. It is our first approximation to the hooked rug S_∞. Let S_1 be the union of disks $E_1, E_2, \ldots, E_n$ such that their interiors are mutually disjoint and adjacent E_i's have an edge in common (as do E_1 and E_n). Experimentation with subdivisions of S_1 shows that we can pick E's whose diameters are small.

We thicken S_1 so that each E_i becomes a topological cube with E_i as center section. We add an eye bolt H_i to the thickened E_i so as to form a cube-with-eye-bolt T_i as shown in Figure IV.10.A. To avoid complicating the figure we do not show certain T_i in front and in back. This T_i is topologically a solid torus and the loop in each H_i loops around the stem of the next H_i (it being understood also that 1 follows n). Then $W_1 = \bigcup T_i$ is a hollow ball with handles. It may be shown that the simple closed curve J shown in Figure IV.10.A cannot be shrunk to a point in the complement of W_1.

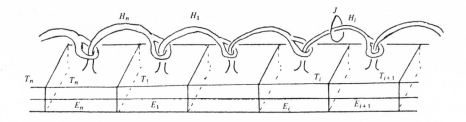

FIGURE IV.10.A

This might be an appropriate time to note that if H_i is unhooked from the stem of H_{i+1}, J could be slipped over the right end of H_i and shrunk to a point. In fact, if instead of unhooking H_i from the stem of H_{i+1}, H_{i+1} were unhooked from the stem of H_{i+2}, J could be shrunk in a more complicated way. A part of it would need to be slipped through the hole of H_i before it finally is pulled over the right end of H_{i+1}. Indeed, if any H_j whatsoever were unhooked from the stem of its H_{j+1}, then with enough maneuvering, J could be slipped off. While we do not need this observation in the description of this particular hooked rug, the observation is useful in building related hooked rugs.

The approximation S_2 will be obtained by replacing each disk E_i in each T_i by another disk in Bd $E_i \cup$ Int T_i so that this replaced disk has the same boundary as E_i. A slice is removed from the loop of each T_i to change T_i into a topological cube K_i. Then Bd E_i separates Bd K_i into two disks one of which has a hook on it. The interior of this hooked disk is pushed slightly into Int T_i so as to form a disk

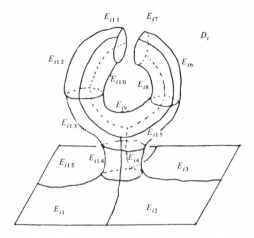

FIGURE IV.10.B

D_i as shown in Figure IV.10.B. The disk, except for its boundary Bd D_i = Bd E_i lies in Int T_i. The 2-sphere $S_2 = \cup D_i$ is the second approximation of S_∞.

Each disk D_i is given a subdivision into small disks $E_{i1}, E_{i2}, \ldots, E_{i15}$ as shown in Figure IV.10.B such that adjacent E_{ij}'s in the sequence share an edge. It is not essential that there be exactly 15 elements in the subdivision and the scheme suggested is only for convenience.

Each E_{ij} is thickened and an eye bolt added to it to form a cube-with-eye-bolt T_{ij}. The loop in the eye bolt T_{ij} goes around the stem of the next eye bolt as shown in Figure IV.10.C. Also, the stems of T_{i7} and T_{i10} intertwine as shown in that figure so that a simple closed curve which cannot be shrunk to a point in the complement of T_i, cannot be shrunk to a point in the complement of $\cup T_{ij}$. The eye bolts run so close to D_i that $\cup T_{ij} \subset T_i$. Then $W_2 = \cup T_{ij}$ is a cube with handles and lies in W_1.

A disk D_{ij} is placed in each T_{ij} in the same fashion that a D_i was placed in each T_i and we define $S_3 = \cup D_{ij}$. Perhaps we have gone far enough for the reader to anticipate the description of S_∞. It is the limit of $S_1, S_2, S_3, \ldots$, and we have already described S_1, S_2, S_3. It is also the intersection of a decreasing sequence of 3-manifolds with boundaries $W_1, W_2, W_3, \ldots$, of which we have already described W_1, W_2.

To get W_{n+1} from S_{n+1} and W_n, each of the special disks D in S_{n+1} is subdivided into 15 small subdisks, each of the special subdisks is thickened into a topological cube and a topological eye bolt is added as suggested in Figure IV.10.C. The cube-with-eye-bolts T shown in Figure IV.10.C are solid even though the figure shows the eye bolts as linear and the thickened subdisks as thin. The 3-manifold-with-boundary W_{n+1} is the sum of the T's at the $(n + 1)$st stage and a disk D like that shown in Figure IV.10.B is put in each sliced T. The sum of these D's is the 2-sphere S_{n+2}. It is understood that as n increases without limit, the mesh of the T's in W_n approaches 0.

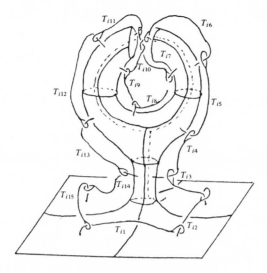

FIGURE IV.10.C

For details of a proof that the limit of the S_i's $= W_1 \cap W_2 \cap \cdots$, is a wild 2-sphere such that each arc in it is tame, the reader is referred to [$\mathbf{B}_{13}$].

While we will not pursue a further study of S_∞ here, we list several of its properties.

> Each arc in S_∞ is tame.
>
> $S_\infty \cup \text{Int } S_\infty$ is a 3-ball.
>
> Ext S_∞ is not simply-connected.
>
> Ext S_∞ is not 1-LC at any point of S_∞.

We mention two variations of S_∞ that have been used to illustrate geometric pecularities.

Instead of having H_{i15} hook the stem of H_{i1}, have it hook the stem of $H_{(i+1)1}$. The resulting 2-sphere S'_∞ has the additional property that if D is any disk in it,

$$D \cup \text{Ext } S'_\infty \text{ is simply-connected.}$$

In another variation of this variation, at each stage we let the last eyebolt dangle instead of having it hook the stem of the first H at this stage. This causes Ext S'_∞ to become simply-connected. See examples of Alford [$\mathbf{A}_6$] and Gillman [$\mathbf{G}_2$].

CHAPTER V

THE GENERALIZED SCHOENFLIES THEOREM

V.1. The Schoenflies theorem for a 2-sphere. The Schoenflies theorem is one of the most interesting and useful theorems from plane topology. It implies that each simple closed curve in R^2 is the boundary of a disk. It is sometimes useful to consider the Schoenflies theorem on a sphere rather than in Euclidean space. If the planar version were shifted to a 2-sphere, it would say that if J is a simple closed curve on a topological 2-sphere, then the 2-sphere is the union of two disks D_1, D_2 such that $D_1 \cap D_2 = J = \text{Bd } D_1 = \text{Bd } D_2$.

Some people have the mistaken idea that since the plane was studied in high school geometry and we know so much about the plane that its study is simpler than the study of higher-dimensional topology. One of the ways to make a study easy is to restrict our attention to easy theorems. This is a way that we sometimes proceed in studying high-dimensional spaces. We are likely to attack very formidable theorems in plane topology and their proofs may be quite involved. It may be noted that the proof of the higher-dimensional Schoenflies theorem that we treat in the next §V.2 of this chapter is much easier than that given in Chapter III. However, in certain respects the theorem treated here is weaker since the hypothesis is more restrictive and the conclusions are weaker.

Early attempts were made to extend the Schoenflies theorem to higher dimensions. A claim [A$_1$] that each topological 2-sphere in R^3 bounds a topological ball was exploded by the 1924 discovery [A$_3$] of the famous Alexander horned sphere described in Chapter IV. The discovery of wild spheres showed that the Schoenflies theorem could not be extended to higher dimensions without modifications. Under what conditions can we conclude that a 2-sphere in R^3 bounds a topological 3-ball. Finding the best set of conditions is still under investigation, but we do have a satisfactory set of conditions that generalizes to all dimensions.

V.2. The generalized Schoenflies theorem. The set of all points in R^{n+1} whose distance from the origin is 1 is called the canonical n-sphere S^n. The canonical $(n-1)$-sphere S^{n-1} in S^n is the set of points of S^n such that the last coordinates of each is 0. Any set homeomorphic to S^n is called a topological n-sphere. The adjective "topological" is usually dropped and we speak of a topological n-sphere as being an n-sphere.

The canonical $(n-1)$-sphere S^{n-1} in S^n has a *cartesian product neighborhood* in that there is a homeomorphism h of $S^{n-1} \times [-1, 1]$ into S^n such that $h(s \times 0) = s$ for each $s \in S^{n-1}$. Brown [B$_{29}$] has called S^{n-1} *bicollared* where each of $h(S^{n-1} \times (-1, 0])$, $h(S^{n-1} \times [0, 1))$ is a *collar*. Others say that S^{n-1} is *flat* in S^n.

Brown [B$_{29}$] showed that a topological $(n-1)$-sphere K^{n-1} in R^n bounds a topological n-ball in R^n if K^{n-1} has a cartesian product neighborhood. This result is called the *generalized Schoenflies theorem*.

Brown's work was encouraged by an interesting discovery by Mazur [M$_5$] that the generalized Schoenflies theorem holds in those cases where the cartesian product neighborhood has a quality of "piecewise linearity" at one of its points. Morse [M$_{15}$] showed how to eliminate Mazur's extra requirement of "piecewise linearity". However, the following proof follows Brown's treatment [B$_{29}$].

THEOREM V.2.A (GENERALIZED SCHOENFLIES THEOREM). *Suppose h is a homeomorphism of S^{n-1} into S^n such that $h(S^{n-1})$ has a cartesian product neighborhood. Then h can be extended to a homeomorphism of S^n onto itself.*

We shall give a proof of this theorem for the case $n = 2$ that applies equally well in higher dimensions. We choose this setting since many readers are better acquainted with a round 2-sphere S^2 with its north pole, south pole, equator, and great circles than they are with corresponding parts of S^n.

PROOF OF GENERALIZED SCHOENFLIES THEOREM IN DIMENSION 2. We suppose S^1 is the equator of S^2, a is the north pole, and b is the south pole. We suppose that in S^2, $h(S^1 \times [-1, 1])$ is a cartesian product neighborhood of $h(S^1)$ and $h(s \times 0) = h(s)$. It follows from the Jordan curve Theorem VII.6.E and invariance of domain Theorem VII.6.F of Chapter VII that $S^2 - h(S^1 \times (-1, 1))$ consists of two components A, B containing $h(S^1 \times 1)$, $h(S^1 \times -1)$ respectively.

The idea for the proof is quite simple. Show that each of B, A is cellular. (A set in S^n is cellular if it is the intersection of a decreasing sequence of n-cells in S^n each of which lies in the interior of the preceding.) Then using the fact that B is cellular show that there is a homeomorphism of the lower half of S^2 (part on or below equator) onto $h(S^1 \times [-1, 0]) \cup B$ such that this homeomorphism agrees with h on S^1. Since A is cellular, a similar agrument shows that there is a homeomorphism of the upper half of S^2 onto $h(S^1 \times [0, 1]) \cup A$.

The part of the proof that requires ingenuity is in showing that each of A, B is cellular. We need to show that for each open set U in S^2 containing B, there is a disk D_U in U such that $B \subset$ Int D_U. We describe D_U as the third member of a sequence of sets D_0, $f_1(D_0)$, $h_0(D_0) = D_U$ in S^2, the first of which is a disk D_0 such that $A \cup B \subset$ Int D_0. We suppose $U \subset D_0$.

Before defining f_1 we select a map f of S^2 onto itself such that

$f(A) = a$,

$f(B) = b$, and

$f(h(s \times t))$ is the point of the semicircle on S^2 from a to b through s whose last coordinate is t.

Then f restricted to $S^2 - (A \cup B)$ is a homeomorphism. We regard f as *killing* A and B. See Figure V.2.A. Let r_1 be a homeomorphism of S^2 onto itself such that $b \notin r_1 f(D_0)$ and r_1 is the identity in some neighborhood of a. Figure V.2.B illustrates such an r_1. Then

$$f_1(x) = \begin{cases} x, & x \in A, \\ f^{-1} r_1 f(x), & x \in D_0 - A. \end{cases}$$

That f_1 is continuous at points of A follows from the restriction that r_1 is the identity in a neighborhood of $f(A) = a$. The map f killed A but f_1 raised it from the dead. Let us consider how $f_1(D_0)$ differs from D_0. It contains A but not B. Although $f_1(B)$ is a point of $f_1(D_0)$, B misses $f_1(D_0)$. We do not yet claim that $f_1(D_0)$ is a disk (although indeed it is). As applied to D_0, the map f_1 more nearly resembles a homeomorphism than f in that f_1 has only one nondegenerate inverse rather than two. We show $f_1(D_0)$ as shaded in Figure V.2.C. It may be shown that $f_1(U)$ is open in S^2 and f_1 is a homeomorphism on $D_0 - B$. We have raised A from the dead. In the following step we raise B from the dead also.

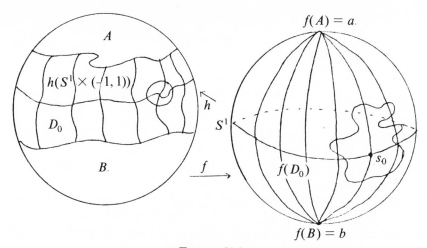

FIGURE V.2.A

We show that B is cellular by showing that for each open subset U of Int D_0 which contains B, there is a homeomorphism h_0 of D_0 into U such that $B \subset$ Int $h_0(D_0)$. Then $D_U = h_0(D_0)$.

Let r_2 be a homeomorphism of S^2 onto itself such that

r_2 is the identity in some neighborhood of the point $f_1(B)$, and

$r_2 f_1(D_0) \subset f_1(U)$.

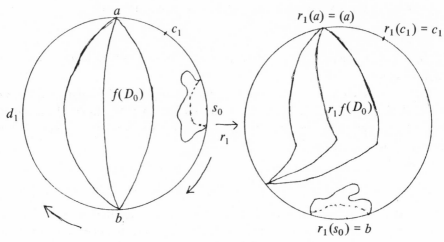

FIGURE V.2.B

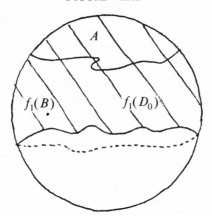

FIGURE V.2.C

Then the required homeomorphism h_0 is defined as

$$h_0(x) = \begin{cases} x, & x \in B, \\ f_1^{-1} r_2 f_1(x), & x \in D_0 - B. \end{cases}$$

We shall show that there is a homeomorphism of the lower half of S^2 onto $h(S^1 \times [-1, 0]) \cup B$ that agrees with h on S^1. Since B is cellular, there is a sequence of topological disks $D_1, D_2, \ldots,$ in S^2 such that

$$D_1 \subset h(S^1 \times [-1, 0)) \cup B,$$
$$\cap D_i = B, \quad \text{and}$$
$$D_{i+1} \subset \text{Int } D_i.$$

Let $\phi_1, \phi_2, \ldots,$ be a sequence of homeomorphisms of S^2 onto itself such that

$$\phi_1 \text{ is the identity on } S^2 - D_1,$$
$$\text{diameter } \phi_1(D_2) \le 1/2^1,$$

$$\phi_2 = \phi_1 \quad \text{on } S^2 - D_2,$$
$$\text{diameter } \phi_2(D_3) \leq 1/2^2,$$

$$\cdots$$

$$\phi_{i+1} = \phi_1 \quad \text{on } S^2 - D_i,$$
$$\text{diameter } \phi_{i+1}^{\cdot}(D_{i+2}) \leq 1/2^{i+1}$$

$$\cdots .$$

Let $\phi = \text{limit } \phi_i$.

A homeomorphism of the lower half of S^2 onto $h(S^1 \times [-1, 1]) \cup B$ is given by ϕf^{-1}. Note that ϕf^{-1} agrees with h on S^1 since $f^{-1} = h$ on S^1 and ϕ is the identity on $h(S^1)$. □

THEOREM V.2.B. *An $(n - 1)$-sphere in S^n (or R^n) is flat if it has a cartesian product neighborhood.* □

The generalized Schoenflies theorem has the following application.

THEOREM V.2.C. *A compact n-manifold is a topological n-sphere if it is the union of two open n-balls.*

PROOF. Suppose the compact n-manifold M^n is the union of two open balls $\mathring{B}_1^n$, $\mathring{B}_2^n$. The compact set $M^n - \mathring{B}_1^n$ lies in $\mathring{B}_2^n$ and there is a topological n-ball B_3^n such that $M^n - \mathring{B}_1^n \subset \text{Int } B_3^n \subset B_3^n \subset \mathring{B}_2^n$ and Bd B_3^n has a cartesian product neighborhood. Since Bd B_3^n also has a cartesian product neighborhood in $\mathring{B}_1^n$, it bounds an n-ball B_4^n in $\mathring{B}_1^n$. Then M^n is a topological n-sphere since it is the union of B_3^n and B_4^n sewed together with the identity homeomorphism on their boundaries. □

A subset X of Y is *collared* in Y if there is a homeomorphism h of $X \times [0, 1)$ into an open subset of Y such that $h(x \times 0) = x$ for each $x \in X$. We call h: $X \times [0, 1)$ a *collar* of X. Also, X is *locally collared* if for each $x \in X$, there is an open subset U of X containing x such that U is collared in Y. The generalized Schoenflies theorem has the following further application.

THEOREM V.2.D. *Suppose K^{n-1} is a topological $(n - 1)$-sphere in R^n and U is a bounded component of $R^n - K^{n-1}$. Then $K^{n-1} \cup U$ is a topological n-ball if and only if K^{n-1} is collared in $K^{n-1} \cup U$.*

PROOF. If $K^{n-1} \times [0, 1)$ is the collar, it follows from Theorem V.2.A that the part of U bounded by $K^{n-1} \times 1/2$ is an n-ball. Then $U \cup K^{n-1}$ is this n-ball with a collar. □

If f is a map of X into Y, the *set of singularities* of f is the closure in X of $\{x \in X: x \neq f^{-1}f(x)\}$. We denote this set of singularities by $S(f)$. The technique we have used to reduce singularities by raising sets from the dead may be worth formulating as a theorem.

THEOREM V.2.E. *Suppose f is a map of compact metric space X into Y, U is an open subset of $f(X)$, $A = U \cap f(S(f)) \neq 0$, and r is a homeomorphism of $f(X)$ into U such that r is the identity on an open set containing A. Let*

$$g(x) = \begin{pmatrix} x & \text{if } x \in f^{-1}(A), \\ f^{-1}rf(x) & \text{if } x \in X - f^{-1}(A) \end{pmatrix}.$$

Then g is continuous and $S(g) = S(f) - f^{-1}(A)$.

PROOF. Both X and $S(f)$ are compact. If $x \in X - S(f)$ there is a neighborhood N of x such that $\overline{N} \cap S(f) = 0$. One finds successively that f, rf, and $f^{-1}rf$ are homeomorphisms on $\overline{N}$. Also, if $x \in f^{-1}(A)$, $f^{-1}rf$ is the identity in some neighborhood of x. Finally, if $x \in S(f) - f^{-1}(A)$, rf is continuous at $rf(x) \in U - fS(f)$. Hence f^{-1} is a homeomorphism in a neighborhood of $rf(x)$. □

EXERCISE V.2.F. Show that the above theorem is false if one omits the hypothesis that X is compact.

Answer. If f is a one-to-one map of a ray onto a circle and r is a rotation of the circle, $f^{-1}rf$ is not continuous at the inverse of the starting point of the ray. □

V.3. The canonical collared Schoenflies theorem. Suppose S_1, S_2 are two flat $(n-1)$-spheres in S^n that are homeomorphically close (there is a homeomorphism of S_1 onto S_2 which does not move any point far). One might wonder if this homomorphism can be extended to a homeomorphism of S^n onto itself that does not move any point of S^n far. A bit of reflection shows that if ε, δ are given (to measure "not far" and "close") before S_1, S_2, it is possible to select S_1, S_2 such that there is a δ-homeomorphism taking one onto the other but there is not ε-homeomorphism taking the ball bounded by one onto the ball bounded by the other. Figure V.3 shows two simple closed curves J_1, J_2 in R^2 (or S^2) that are homeomorphically within δ of each other but the disks bounded by them are not ε close. The purpose of these remarks was to make Theorem V.3.A appear somewhat surprising. We shall be interested in conditions under which we can get a rule for extending a homeomorphism h of an $(n-1)$-sphere in S^n into S^n to a homeomorphism of S^n onto itself.

If X, Y are metric spaces with X compact, let $I(X, Y)$ *denote the set of all homeomorphism $\{h_\alpha\}$ of X into Y metrized by*

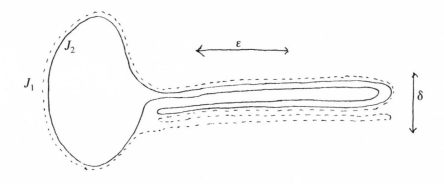

FIGURE V.3

$$d(h_\beta, h_\gamma) = \max_{x \in X} \ d(h_\beta(x), h_\gamma(x)).$$

This notation provides a vehicle for stating the following theorem.

THEOREM V.3.A (CANONICAL COLLARED SCHOENFLIES THEOREM FOR S^n). *There is a continuous map* Φ *of* $I(S^{n-1} \times [-1, 1], S^n)$ *into* $I(S^n, S^n)$ *such that each* $\phi(h_\alpha) \mid S^{n-1} \times 0 = h_\alpha \mid S^{n-1} \times 0.$

Perhaps the most spectacular application of the canonical collared Schoenflies theorem was at the third level of Kirby's remarkable diagram [**K₂**] giving certain properties of maps of R^n onto itself. He used it to show that if $T^n = S^1 \times S^1 \times \cdots \times S^1$ is an n-torus and B_1, B_2 are n-cells on T^n with $B_1 \subset \text{Int } B_2$, then if h_1, h_2 are nearby elements of $I(T^n - \text{Int } B_1, T^n)$ then there are nearby elements g_1, g_2 of $I(T^n, T^n)$ such that g_i agrees with h_i on $T^n - B_2$.

We give a proof in dimension 2 that works equally well in higher dimensions. The proof consists of showing that the sets and maps used in the proof of the generalized Schoenflies theorem given in the last section can be described in a continuous canonical fashion. The proof is complicated and the reader who is only skimming may choose to skip to §V.4 and temporarily forego the rigors of the proof.

PROOF OF CANONICAL COLLARED SCHOENFLIES THEOREM FOR S^2. We seek a canonical way for describing the maps and sets used in the proof of Theorem V.2.A. The map $h \colon S^1 \times [-1, 1]$ is given, and A, B are precisley defined. Also, f is canonical. Instead of defining D_0 to be any disk in S^2 with $A \cup B \subset \text{Int } D_0$ we pick it to be a special one as follows. Let s_0 be the point of S^1 whose first coordinate is 1 and let D_0 be the large disk which is the complement of the small open disk with center at $h(s_0 \times 0)$ and radius $\frac{1}{3}d(h(s_0 \times 0), A \cup B))$. This makes the definition of D_0 specific and if two h's are close, their associated D_0's are close.

To pick f_1 to be canonical, we need to select an r_1 that is. Figure V.2.B shows r_1 is a variable rotation. If one wished an equation of such a rotation, one might consider the great circle on S^2 through a, s_0, and b. Let c_1, d_1 be the points of this great circle in the order a, c_1, s_0, b, d_1, a where c_1 is halfway between a and s_0 while d_1 is halfway between b and a. Then restricted to the great circle, r_1 is the identity on d_1ac_1, takes c_1s_0 linearly onto c_1b, and takes s_0d_1 linearly onto bd_1. Also, r_1 is defined to be the identity on the poles of S^2 with respect to this great circle. Elsewhere on S^2, r_1 is defined to take each semicircle on S^2 joining these poles linearly only another such semicircle. The second such semicircle is determined by the image of the midpoint of the first.

We proceed in this fashion to select r_1, h_0, D_U, the D_i's, and ϕ_i' canonically. This makes the homeomorphism ϕf^{-1} canonical. □

Question. If B^3 is a 3-ball, it is not possible to get a map $\Phi \colon I(\text{Bd } B^3, R^3) \to I(B^3, R^3)$ such that for each $h_\alpha \in I(\text{Bd } B^3, R^3)$, Φh_α agrees with h_α on Bd B^3

since some $h_\alpha(\text{Bd } B^3)$'s are wild from their interiors. If $I'(\text{Bd } B^3, R^3)$ is the set of elements of $I(\text{Bd } B^3, R^3)$ whose images are tame, is there a map $\Phi: I'(\text{Bd } B^3, R^3) \to I(B^3, R^3)$ such that Φh_α and h_α agree on Bd B^3? Our proof of the canonical collared Schoenflies theorem made use of the given collar but one might think that it might be enough to know that there be a collar rather than have one selected.

Theorem V.3.A shows that it is possible to map one topological space into another subject to certain restrictions. This is a type of selection theorem. Some selection problems are quite hard. We give four examples of questions of which we know only the answers to the first and fourth. In each case we are dealing with a collection of homeomorphic compact objects in the same space and decide that the distance between two objects is the least upper bound of all ε's such that there is an ε-homeomorphism of one object onto the other.

1. If G is a collection of straight line segments in R^2, is it possible to continuously select a point from each?

2. If G is the collection of all arcs in R^2, is there a map $f: G \to R^2$ such that for each $g \in G, f(g) \in g$?

3. If G is the collection of all simple closed curves in R^2, is there such a selection map $f: G \to R^2$?

4. If G is the collection of all simple closed curves on S^2, is there such a selection map $f: G \to S^2$?

To see that the answer to Question 4 is negative, one might restrict his attention to the circles of radius $\frac{1}{4}$ on the round unit S^2. If a point is selected from a circle, the point determines a vector on S^2 from the center of the small disk bounded by the circle to the selected point. That one cannot get a continuous collection of vectors on a 2-sphere is sometimes interpreted by the layman to mean that there is a place on the earth where the wind is not blowing—you cannot comb the hair on a billiard ball.

This does not answer the question as to whether there is a map of $I'(S^{n-1}, S^n)$ into $I(S^n, S^n)$ such that a homeomorphism in $I'(S^{n-1}, S^n)$ and its images in $I(S^n, S^n)$ agree on S^{n-1}.

V.4. Local flatness. Suppose M^{n-1} is a topological $(n-1)$-manifold in an n-manifold N^n. We say that M^{n-1} is *locally flat* if for each point $p \in M^{n-1}$ there is a neighborhood U in N^n of p and a homeomorphism of U into R^n that takes $U \cap M^{n-1}$ onto a canonical $(n-1)$-plane in R^n. Locally having a cartesian product neighborhood or being locally bicollared are essentially the same as being locally flat although these definitions use the fiber approach rather than the hyperplane one. Other authors use fibers with dimensions greater than or equal to 1 (see [$\mathbf{R_1}$]), but we shall not. We learn from Theorem V.4.A that S^{n-1} is tame if it is locally flat.

THEOREM V.4.A. *Any homeomorphism of one locally flat $(n-1)$-sphere in S^n (or R^n) onto another in S^n (or R^n) can be extended to a homeomorphism of S^n (or R^n) onto itself.*

The theorem follows from Theorems V.2.A and V.2.B when it is shown that an $(n-1)$-sphere has a caretsian product neighborhood if it is locally flat. □

THEOREM V.4.B. *Suppose the connected $(n-1)$-manifold M^{n-1} in the n-manifold N^n is the common boundary of two components of $N^n - M^{n-1}$. Then M^{n-1} has a cartesian product neighborhood in N^n if M^{n-1} is locally flat.*

Some two-sidedness condition in the hypothesis of Theorem V.4.B is required. There is a simple closed curve J in a projective plane P^2 that does not separate P^2. Although J is locally flat in P^2, it is the center line of a Moebius band and does not have a cartesian product neighborhood.

In proving Theorem V.4.B, it is convenient to work with a component U of $N^n - M^{n-1}$ and note that M^{n-1} is locally collared in $M^{n-1} \cup U$. We shall find that the proof that M^{n-1} is collared in $M^{n-1} \cup U$ makes no use of the manifold properties of M^{n-1} so we state the results as follows. □

THEOREM V.4.C. *A subset X of metric space Y is collared in Y if it is locally collared in Y.* □

The following two lemmas are used in the proof of Theorem V.4.C. They show how one may adjust a collar. The adjustment was suggested by Brown in [**B₃₀**]. A shorter version was given by Connerly in [**C₁₀**].

LEMMA V.4.D. *Suppose U_1, U_2 are open subsets of X with $U_2 \subset U_1$ and h_1: $U_1 \times [0,1)$, h_2: $U_2 \times [0,1)$ are collars in Y. Then for each open set V in Y containing U_2 there is an open set V_1 in Y and a collar h: $U_1 \times [0,1)$ such that*

$$U_2 \subset V_1 \subset V,$$
$$h = h_1 \quad on \ h_1^{-1}(Y - V), \quad and$$
$$h = h_2 \quad on \ h_2^{-1}(V_1).$$

The lemma is proved by raising a blister over U_2. In describing blisters we shall consider certain maps f of U_2 into $(0,1]$ and use $h_i(U_2 \times [0, f))$ to denote the variable cartesian product $\bigcup_{x \in U_2} h_i(x \times [0, f(x)))$.

PROOF OF LEMMA V.4.D. Let f_1, f_2 be two maps of U_2 into $(0,1]$ such that

$h_1(x \times t) \in V$ and is nearer to x than to $X - U_2$ if $x \in U_2$ and $t \leq f_1(x)$,
$f_2 < f_1$,
$h_2(U_2 \times [0, f_2)) \subset h_1(U_2 \times [0, f_1))$.

The required open set V_1 is the blister $h_2(U_2 \times [0, f_2/2))$. In a certain sense $V_2 = h_2(U_2 \times [0, f_2))$ is a blister twice as high. See Figure V.4.D.

Let g be the homeomorphism of Y into itself that is the identity on $Y - V_2$ and for each $x \in U_2$ sends $h_2(x \times t)$ to $h_2(x \times (t + f_2(x))/2)$ for $0 \leq t \leq f_2(x)$. In a certain sense, g raises a blister over U_2 and squeezes V_2 into part of V_2 above V_1. The homeomorphism frees the blister V_1 since it sends $h_2(x \times [0, f_2(x)))$ onto $h_2(x \times [f_2(x)/2, f_2(x)))$. If $A = x \times [0, f_2(x))$, Figure V.4.D shows $h_2(A)$, $gh_2(A)$, $h_1(A)$, and $gh_1(A)$. Points are raised.

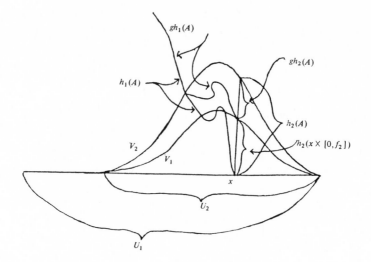

FIGURE V.4.D

The required homeomorphism $h\colon U_1 \times [0, 1)$ sends $x \times [0, 1)$ onto

$$h_2(x \times [0, f_2/2)) \cup gh_1(x \times [0, 1))$$

where the parametrization is adjusted along $h(x \times [0, 1))$ in a natural way as follows. If r_x is the linear map that sends the segment $[f_2(x)/2, f_2(x)]$ onto $[0, f_2(x)]$,

$$h(x \times t) = h_1(x \times t) \quad \text{for } 0 \leqslant t \leqslant f_2(x)/2,$$
$$h(x \times t) = gh_2(x \times r_x(t)) \quad \text{for } f_2(x)/2 \leqslant t \leqslant f_2(x),$$
$$h(x \times t) = gh_2(x \times t) \quad \text{for } f_2(x) \leqslant t < 1. \quad \square$$

LEMMA V.4.E. *Suppose U_1, U_2 are open subsets of X and $h_1\colon U_1 \times [0, 1)$, $h_2\colon U_2 \times [0, 1)$ are collars in Y such that each $h_1(x \times t)$ is nearer to x than to $X - U_1$. Then there is a collar $h\colon (U_1 \cup U_2) \times [0, 1)$ such that $h = h_1$ on $(U_1 - U_2) \times [0, 1)$ and each $h(x \times t)$ is nearer to x than to $X - (U_1 \cup U_2)$.*

PROOF. Since Lemma V.4.D says that we can adjust h_2 to make it agree with h_1 in a neighborhood of $(U_1 \cap U_2) \times 0$ in $(U_1 \cap U_2) \times [0, 1)$, we suppose with no loss of generality that there is a map $f_1\colon (U_1 \cap U_2) \to (0, 1)$ such that h_2 agrees with h_1 on $(U_1 \cap U_2) \times [0, f_1)$.

Let U_1', U_2' be disjoint open subsets of $U_1 \cup U_2$ containing $U_1 - U_2$, $U_2 - U_1$ respectively. Since each $h_1(x \times t)$ is closer to x than to $X - U_1$, there is a map $f_2\colon U_2 \to (0, 1)$ such that $h_1(U_1' \times [0, 1))$ misses $h_2(U_2' \times [0, f_2))$ and for $t < f_2(x)$, $h_2(x \times t)$ is nearer to x than to $X - U_2$. Let f be a map of $U_1 \cup U_2$ into $(0, 1)$ such that

$$f = 1 \quad \text{on } U_1 - U_2,$$
$$f < f_1 \quad \text{on } U_1 \cap U_2 - (U_1' \cup U_2'),$$
$$f < f_2 \quad \text{on } U_2'.$$

Letting r_x be the linear map that takes $[0, 1]$ to $[0, f(x)]$, we define h as follows

$$h(x \times t) = h_1(x \times r_x(t)) \quad \text{for } x \in U_1 - U_2',$$
$$h(x \times t) = h_2(x \times r_x(t)) \quad \text{for } x \in U_2 - U_1'.$$

It is to be noted that $h_1 = h_2$ on $x \times r_x(t)$ for $x \in (U_1 \cap U_2) - (U_1' \cup U_2')$ since $f < f_1$. $\square$

PROOF OF THEOREM V.4.C. Let $\{U_\alpha\}$ be an open covering of X such that each U_α has a collar in Y. Since a metric space is paracompact, we can suppose with no loss of generality that $\{U_\alpha\}$ is locally finite—that is, each point of X lies in an open subset U of X such that U intersects at most a finite number of elements of $\{U_\alpha\}$. For each U_α let $h_\alpha(U_\alpha \times [0, 1))$ be a collar for U_α such that each $h_\alpha(x \times t)$ is closer to x than to $X - U_\alpha$.

Theorem V.4.C follows directly from Lemma V.4.E if $\{U_\alpha\}$ is finite or countable but requires a bit of fancy footwork if $\{U_\alpha\}$ is uncountable.

Suppose $\{U_\alpha\}$ is uncountable and $W = (1, 2, \ldots, \alpha, \ldots)$ is a well ordering with no last element of the index set of $\{U_\alpha\}$. For each $\beta \in W$ we let $U_\beta^+ = \bigcup_{\alpha < \beta} U_\alpha$. As indicated in the following paragraph, it follows from Lemma V.4.E that there is an uncountable sequence of collars $h_2: U_2^+ \times [0, 1)$, $h_3: U_3^+ \times [0, 1), \ldots, h_\alpha: U_\alpha^+ \times [0, 1), \ldots,$ such that each $h_\alpha(x \times t)$ is nearer to x than to $X - U_\alpha^+$ and if $\beta_1 < \beta_2$ and $x \in U_{\beta_1}^+$ but $x \notin \bigcup_{\beta_1 = \alpha < \beta_2} U_\alpha$, then $h_{\beta_1}(x \times t) = h_{\beta_2}(x \times t)$. To define h on $x \times [0, 1)$ we let U_γ be the last element of $\{U_\alpha\}$ to contain x and let $h = h_{\gamma+1}$ on $x \times [0, 1)$.

We shall abbreviate $h_\alpha: U_\alpha^+ \times [0, 1)$ by h_α. To prove that there is an uncountable sequence $h_2, h_3, \ldots, h_\alpha, \ldots,$ as mentioned in the preceding paragraph, one might proceed as follows. For each U_α^+ for which there is a collar h_α such that each $h_\alpha(x \times t)$ is closer to x than to $X - U_\alpha^+$, let W_α be a well ordering of all such collars. For two sequences $h_2^i, h_3^i, \ldots, h_\alpha^i, \ldots$ $(i = 1, 2)$ we use the lexicographical ordering to see which sequence comes first—find the first place where the sequences differ (say in the γth slot) and then decide which of the two sequences comes first on the basis of whether or not h_γ^1 or h_γ^2 comes first in W_γ: if the two sequences agree on each place where each is defined, the shorter sequence comes first. Instead of showing that there exists a sequence $h_2, h_3, \ldots, h_\alpha, \ldots,$ one has the easier task of showing that there is a unique such sequence with the additional property that $h_{\alpha+1}$ is the first sequence in $W_{\alpha+1}$ that agrees with h_α on $U_\alpha^+ - U_\alpha$. If α is a limit ordinal, h_α is automatically determined by the nature of its predecessors.

V.5. Weakness of the generalized Schoenflies theorem. A homeomorphism of R^n onto itself is *stable* if it is the composition of a finite number of homeomorphisms each of which is fixed on an open set. The Schoenflies homeomorphisms given by Theorems III.1.A and III.6.C are stable since they have compact support. A translation of Theorem V.2.A to R^n would say that if B_1^n, B_2^n are

n-balls with locally flat boundaries in R^n, then there is a homeomorphism h: $R^n \to R^n$ such that $h(B_1^n) = B_2^n$. However there is no assurance that h does not shake R^n at its very roots.

In Chapter XIV we do show that in the case where $n = 3$ and the Bd B_i^n are PL then there is a PL homeomorphism h: $R^3 \to R^3$ such that $h(B_1^n) = h(B_2^n)$ and h has compact support. The side approximation theorem for R^3 is given in Chapter XIII. As pointed out there, this result (or even the approximation theorem), can be used to show that if M^2 is a locally flat 2-manifold in R^3 and $\varepsilon > 0$, there is a homeomorphism h: $R^3 \to R^3$ such that $d(h, \text{Identity}) < \varepsilon$, h is identity outside ε-neighborhood of M^2, and $h(M^2)$ is PL. Hence if B_1^3, B_2^3 are two 3-balls with locally flat boundaries in R^3 there is a homeomorphism h: $R^3 \to R^3$ with compact support such that $h_1(B_1^3) = h_1(B_2^3)$. However, it is to be noted that the method of proof is more difficult than that used in Theorem V.2.A.

The *annulus conjecture* is that if B_1^n, B_2^n are locally flat n-balls in R^n with $B_1^n \subset \text{Int } B_2^n$, $B_2^n - \text{Int } B_1^n$ is homeomorphic to $S^{n-1} \times [-1, 1]$. Theorem V.2.A does not provide a proof of the annulus conjecture since the homeomorphism h given, might not be fixed on any open set. However the techniques of Chapter III and the applications cited in Chapter XVIII can be used to establish the annulus conjecture in R^2 and R^3.

The techniques used to prove Theorem V.2.A about a locally flat $(n - 1)$-sphere in R^n does not seem to extend to apply to locally flat $(n - 1)$-manifolds. They apply only to $(n - 1)$-spheres. However the approximation theorems given in Chapter XIII do apply to 2-manifolds in R^3.

CHAPTER VI

THE FUNDAMENTAL GROUP

In Chapter VI we had need to show that certain sets are not simply-connected. In the present chapter we introduce machinery that is frequently used for these purposes. We give an algorithm for computing the fundamental group of the complement of a finite graph in R^3. The study of fundamental groups is a part of homotopy theory. However, it is not our aim to consider the broad and useful area of homotopy theory but rather to concentrate on a part of it that has had such a strong impact on the study of 3-manifolds.

Some authors regard R^3 as euclidean and denote it by E^3. To familiarize the reader with this notation, we shall occasionally use it in this chapter.

VI.1. Paths and loops. A *path* in a topological space X is a map f of the segment $[0, 1]$ into X. The images $f(0)$ and $f(1)$ are called the *beginning* and *end points* of the path. They are also called the *initial* and *terminal points*.

Multiplication of paths. If f_1, f_2 are paths in X such that the end of f_1 is the beginning of f_2, the *product of f_1 and f_2* (written $f_1 * f_2$) is the path

$$f_1 * f_2(t) = f_1(2t), \qquad 0 \leqslant t \leqslant \tfrac{1}{2},$$
$$= f_2(2t - 1), \qquad \tfrac{1}{2} \leqslant t \leqslant 1.$$

In moving along the path $f_1 * f_2$ as suggested in Figure VI.1.A, one moves along the path f_1 at double speed until the end is reached and then moves along f_2 at double speed. It is to be noted that $f_1 * f_2$ is not defined unless the end of f_1 is the beginning of f_2.

Inverse. If f is a path, we call

$$f^{-1}(t) = f(1 - t), \qquad 0 \leqslant t \leqslant 1,$$

the *inverse* of f. As indicated in Figure VI.1.B, we are moving along f in the opposite direction.

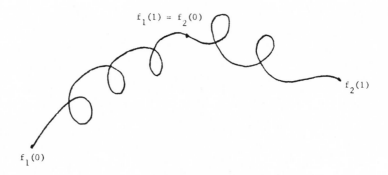

FIGURE VI.1.A

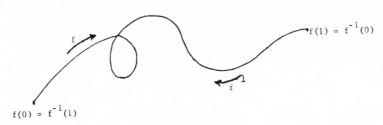

FIGURE VI.1.B

Homotopy of paths. If f_1, f_2 are two paths with the same beginnings and ends we say that f_1 *is homotopic to* f_2 *with ends fixed* (written $f_1 \sim f_2$) if there is a continuous family of maps $F_t : [0, 1] \to X \,|\, (0 \leq t \leq 1)$ such that $F_0 = f_1$, $F_1 = f_2$, $F_t(0) = f_1(0) = f_2(0)$, $F_t(1) = f_1(1) = f_2(1)$. The map is sometimes called a *homotopy*. It may be convenient to regard the homotopy as a continuous shifting of the path f_1 until it becomes the path f_2 as suggested in Figure VI.1.C. We may regard F as mapping the unit planar square with the opposite vertices $(0, 0)$, $(1, 1)$ into X such that $F(x, y) = F_y(x)$. Note that F is constant on each vertical side of the square. On the bottom of the square F is determined by f_1, and on the top it is determined by f_2.

Equivalence classes of paths. The above homotopy breaks the paths in X into equivalence classes. We use $[f]$ to denote the set of all paths g in X such that $f \sim g$. To see that $f \sim g$ implies $g \sim f$ one need only reflect the unit square across a horizontal segment through its center. To visualize the fact that $f_1 \sim f_2$ and $f_2 \sim f_3$ implies $f_1 \sim f_3$ one might consider the maps F and G on the unit squares showing that $f_1 \sim f_2$ and $f_2 \sim f_3$ respectively, squash each of these squares to one half its normal height, and then set the second squashed square on top of the first. The homotopy H showing that $f_1 \sim f_3$ is given by

$$H_t = F_{2t}, \qquad 0 \leq t \leq \tfrac{1}{2},$$
$$= G_{2t-1}, \qquad \tfrac{1}{2} \leq t \leq 1.$$

Finally, $f \sim f$.

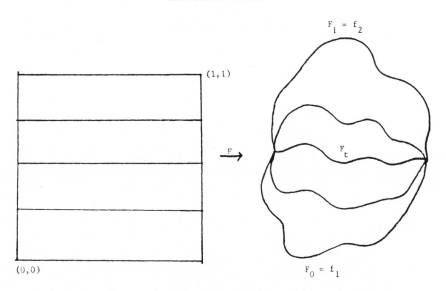

FIGURE VI.1.C

Associativity. Suppose f_1, f_2, f_3 are paths such that the end of f_1 is the beginning of f_{i+1}, $i = 1, 2$. Then $(f_1 * f_2) * f_3$ and $f_1 * (f_2 * f_3)$ are defined, although they may be different as paths. Under $(f_1 * f_2) * f_3$ one considers moving along f_1 at four times normal speed, along f_2 at four times normal speed, and along f_3 at double speed. The rates of speed on f_1 and f_3 are interchanged for $f_1 * (f_2 * f_3)$. If F is defined so as to be constant on each of the slanting segments shown in Figure VI.1.D, there results a homotopy showing that $f_1 * (f_2 * f_3) \sim (f_1 * f_2) * f_3$.

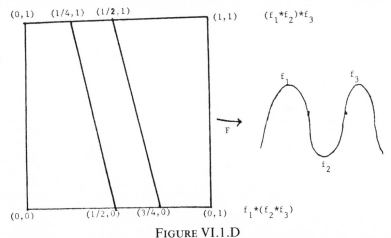

FIGURE VI.1.D

Loops. Select a point $x_0 \in X$ and designate it as a *base point*. A *loop* in X based at x_0 is a path in X that has this base point as both a beginning and an end point. See Figure VI.1.E. Note that any two loops based at x_0 can be multiplied. This puts us in a position to use equivalence classes of loops based at x_0 as elements of a group.

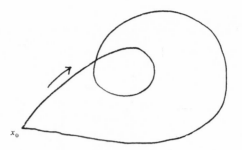

FIGURE VI.1.E

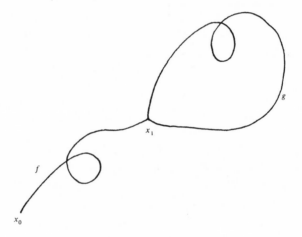

FIGURE VI.2

VI.2. The fundamental group. Let x_0 be a base point in a topological space X. We use $\pi_1(X, x_0)$ to denote the *fundamental group* of X based at x_0. An element of this fundamental group is an equivalence class of loops in X based at x_0. If $[f] \in \pi_1(X, x_0)$, $[f]^{-1}$ is the equivalence class of loops containing f^{-1}. To multiply two elements $[f]$, $[g]$ of $\pi_1(X, x_0)$ we select any two elements $f \in [f]$, $g \in [g]$, and let $[f] * [g] = [f * g]$. It can be shown that this multiplication is independent of the elements chosen in the class. It turns out that the identity element of $\pi_1(X, x_0)$ is the equivalence class containing the constant map $f[0, 1] = x_0$. Multiplication is associative and π_1 is a group.

It may be shown that if there is a path f in X from x_0 to x_1, then $\pi_1(X, x_1)$ is isomorphic to $\pi_1(X, x_0)$. To get such an isomorphism one could let the loop g based at x_1 correspond to the loop $f * g * f^{-1}$ based at x_0, as shown in Figure VI.2. This correspondence between loops is extended to a correspondence between equivalence classes of loops.

When dealing with a pathwise connected space X, it is customary to ignore the base point and use $\pi_1(X)$ to designate $\pi_1(X, x_0)$.. We shall frequently do this in the following sections.

VI.3. Graphs. A PL *finite graph* in E^3 is the sum of a finite number of straight segments such that if two of these segments intersect, the intersection is an end point of each. It is a 1-complex.

A plane P in E^3 is in *regular position* with respect to a PL finite graph G if only a finite number of lines normal to P intersect G in more than one point and each of these exceptional normal lines intersects G in precisely two points which are interior points of the segments used to describe G. We wish to alter situations as shown in the left side of Figure VI.3.A so that they become as shown in the right side.

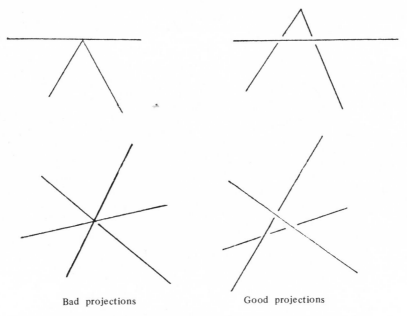

Bad projections   Good projections

FIGURE VI.3.A

If a plane P is in regular position with respect to G, any plane parallel to P is in regular position with respect to G. If one considers the unit vectors from the origin which are normal to planes which are in regular position with respect to G, these vectors intersect the unit sphere with center at the origin in a dense subset of the unit sphere. Hence, there are many vectors such that any plane normal to this vector are in regular position with respect to G.

Consider a projection G' of G into the plane P which is in regular position with respect to G. For convenience we regard P as horizontal so that each point of G is on a vertical line with its corresponding point in G'. If one segment of G contains a point p which is beneath a second segment we say that p is an *undercrossing point* and the first segment *passes beneath* the second. In drawing G', it is convenient to show a gap in the lower segment if one passes beneath the other. In the graph G (or its projection G') shown in Figure VI.3.B, there are five undercrossing points, three branch points and one end point.

Although for technicalities in giving proofs, one may give lip service to the use of polyhedral graphs, in practical situations one usually draws smooth graphs.

The graph shown in Figure VI.3.B is shown as a smooth graph in Figure VI.3.C. In working with smooth graphs one insists that if one smooth segment of G passes underneath another, there is actual crossing in G' and not mere tangency.

There is a theory for studying the fundamental groups of the complements of wild graphs but that is not treated in this book.

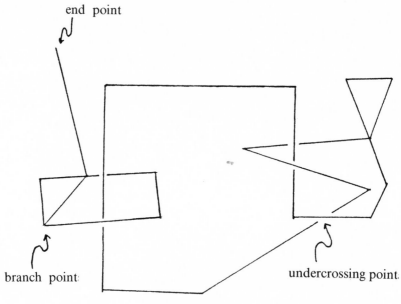

FIGURE VI.3.B

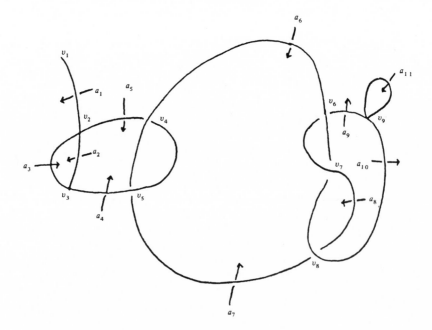

FIGURE VI.3.C

VI.4. Associative words with loops. Suppose G is a graph (either polygonal or smooth) in E^3 with a nice projection as shown in either Figure VI.3.B or VI.3.C. Let $v_1, v_2, \ldots, v_m$ be the points of G that are end points, branch points, or undercrossing points. Let $A_1, A_2, \ldots, A_n$ be the collection of open arcs and simple closed curves of G such that no point of any A_i is a v_j but each boundary point of A_i is a v_j. The closure of each A_i is either an arc in G or a simple closed curve in G which intersects at most one v_j. To each A_i we associate an arrow going under A_i and designate it with a letter a_i as shown in Figure VI.3.C. It does not matter in which direction the arrows (short oriented vectors) point, but it is customary to think of moving along the A_i's of G in some direction and for the arrows to go under from the left side of A_i to the right. In most drawings, there is no need to label the v_i's and the A_i's since it is only the a_i's that are used.

Suppose x_0 is a point of $E^3 - G$ selected as a base point and f is a loop based at x_0 taking $[0, 1]$ into $E^3 - G$. Figure VI.4 shows the image of such an f. We suppose that f is *normal* with respect to G in the sense that $[0, 1]$ can be broken into a finite number of segments $[b_0, b_1], [b_1, b_2], \ldots, [b_{r-1}, b_r]$ $(0 = b_0 < b_1 < \cdots < b_r = 1)$ such that f is a homeomorphism on each $[b_{i-1}, b_i]$ and either no point of $f[b_{i-1}, b_i]$ is beneath G or exactly one point of $f[b_{i-1}, b_i]$ is beneath G and f crosses under a segment of G at this point but not under an undercrossing, branch, or end point of G. Although not each loop of $[f]$ has such nice undercrossing properties, for each $g \in [f]$ and each $\varepsilon > 0$ there is a piecewise linear loop f such that $d[f, g] < \varepsilon$. Pushes can change f to a piecewise linear loop in $[f]$ that is normal with respect to G.

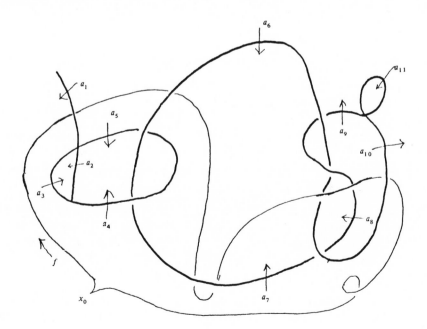

FIGURE VI.4

To each of the r pieces of $[0, 1]$ we associate either a blank or a letter. If $f[b_{i-1}, b_i]$ does not cross under G, we associate a blank. If $f[b_{i-1}, b_i]$ crosses under A_j in the direction of the arrow a_j, we associate the letter a_j; if it crosses under A_j in the opposite direction, we associate a_j^{-1}.

To the loop f we associate a word made up of letters

$$a_1, a_1^{-1}, a_2, a_2^{-1}, \ldots, a_n, a_n^{-1}$$

written as follows: We ignore blanks and first write down the letter associated with $f[b_0, b_1]$, then the letter associated with $f[b_1, b_2], \ldots$, and finally the letter associated with $f[b_{r-1}, b_r]$. In case f crosses under G we obtain a word with a finite number of letters. The word associated with the loop shown in Figure VI.4 is $a_1^{-1} a_6 a_7^{-1} a_7 a_{10}$. It does not contain the letters a_{10}^{-1} and a_8^{-1} since f went over G at these crossings rather than under. (Some authors borrow the convention from functions that if fg is a composite function then g is performed first. They let the last letter correspond to the first undercrossing. We find it convenient not to write backwards and use the nth letter to denote the nth undercrossing.)

Let $W(f)$ be the word associated with f.

Suppose f_1, f_2 are two loops associated with the same word $W(f_1) = W(f_2) = x_1 x_2 \cdots x_s$. It may be shown in this case that $f_1 \sim f_2$ so f_1, f_2 belong to the same element of $\pi_1(E^3 - G)$. This is shown by the following exercise.

EXERCISE VI.4.A. Show that if $W(f_1) = W(f_2)$ then

$$f_1 \sim f_2.$$

Solution. Let $y_1, y_2, \ldots, y_s$ and $y_1', y_2', \ldots, y_s'$ be values of $[0, 1]$ such that the $f_1(y_i)$'s and the $f_2(y_i)$'s are beneath G. Since there is a homeomorphism h of $[0, 1]$ onto itself so that $h(y_i') = y_i$, we suppose without loss of generality that $y_i' = y_i$.

We start modifying f_1 by getting an f_3 such that $W(f_1) = W(f_2) = W(f_3)$ and $f_2(y_i) = f_3(y_i)$. To get f_3, we modify f_1 near the y_i's. As shown in Figure VI.4.A.a. A short arc $p_i q_i$ of $f_1[0, 1]$ about $f_1(y_i)$ is removed, and it is replaced by a folded arc that travels from p_i up vertically to near G, then travels parallel to G (but not crossing over or under G) until it is almost over $f_2(y_i)$, then descending to $f_2[0, 1]$ and crosses under G with an arc of $f_2[0, 1]$ containing $f_2(y_i)$, then rises vertically to near G, then runs parallel to G back to a point above q_i and then finally drops down to q_i. Since $f_3[y_{i-1}, y_i]$ has the same ends as $f_2[y_{i-1}, y_i]$ and each runs above G except at its end points, there is a homotopy of f_3 to f_2 in $E^3 - G$ that leaves each of $0, y_1, \ldots, y_s, 1$ fixed. Since $f_1 \sim f_3 \sim f_2$, $[f_1]$ and $[f_2]$ are the same element of $\pi_1(E^3 - G)$. $\square$

Although we have shown that $f_1 \sim f_2$ if $W(f_1) = W(f_2)$, one should not conclude the converse that $f_1 \sim f_2$ implies $W(f_1) = W(f_2)$. Such is not the case. It may be that f_1 ducks under A_i as shown in Figure VI.4.A.b and picks up the

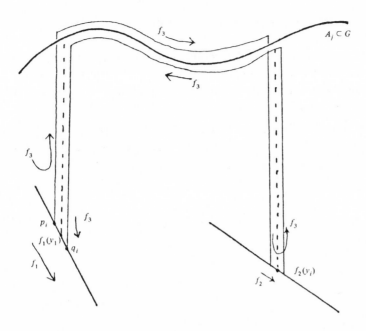

FIGURE VI.4.A.a

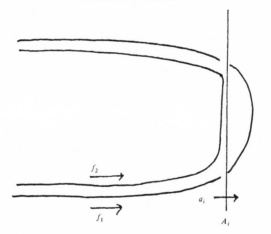

FIGURE VI.4.A.b

letters $a_i a_i^{-1}$. In fact two loops f_1, f_2 are homotopic if $W(f_2)$ can be obtained by either cancelling or inserting the adjacent letters $a_i a_i^{-1}$ or $a_i^{-1} a_i$ any place in $W(f_1)$. In the next section we point out other conditions under which different words are associated with homotopic loops.

VI.5. Relations. We shall consider operations by which $W(f_1)$ can be changed to $W(f_2)$ if $f_1 \sim f_2$ in $E^3 - G$. To do this we consider a relation at each end point, branch point, or undercrossing point of G. A *relation* (designated by $r_i = 1$) is a word r such that it is permitted to cancel or insert either r_i or r_i^{-1} anywhere in any word. (If $r_i = x_1 x_2 \cdots x_s$, $r_i^{-1} = x_s^{-1} x_{s-1}^{-1} \cdots x_1^{-1}$.)

VI.5.A. *End point relation.* If A_j reaches out to v_i as shown in Figure VI.5.A, the loop f going under A_j is homotopic to the loop f' which agrees with f generally except that it goes around the v_i end of A_j instead of crossing under. The relation $r_i = 1$ corresponding to v_i is $a_j = 1$.

VI.5.B. *Branch point relation.* If a loop f goes under the A's reaching out from a branch point as shown in Figure VI.5.B, it is homotopic to the loop f' partially shown. The relation $r_i = 1$ in this case is $a_s^{-1} a_k a_j a_r^{-1} = 1$.

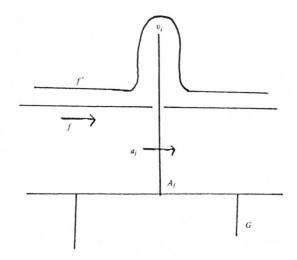

FIGURE VI.5.A

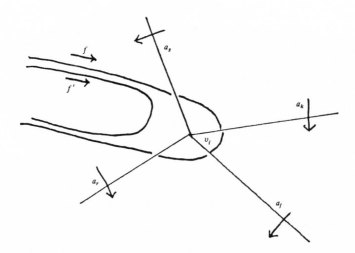

FIGURE VI.5.B

VI.5.C. *Undercrossing relation.* Suppose f goes around an undercrossing as shown in Figure VI.5.C. Then $f' \sim f$. The relation $r_i = 1$ is $a_j a_s a_j^{-1} a_r^{-1} = 1$.

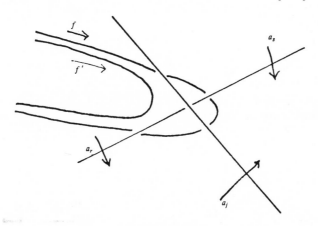

FIGURE VI.5.C

Suppose we associate a relation $r_i = 1$ with each end point, branch point, and undercrossing point of G. Then $f_1 \sim f_2$ in $E^3 - G$ if $W(f_1)$ can be changed to $W(f_2)$ by a finite sequence of the following operations:

(1) Inserting xx^{-1} or $x^{-1}x$ anywhere.

(2) Canceling xx^{-1} or $x^{-1}x$ anywhere.

(3) Inserting any r_i or its inverse anywhere.

(4) Canceling any r_i or its inverse anywhere.

We shall show in the next section that the converse of the preceding paragraph is true in that if $f_1 \sim f_2$ in $E^3 - G$, then $W(f_1)$ can be changed to $W(f_2)$ with operations of these types.

We did not impose definite rules as to how to get the r_i's. For example, one person looking at Figure VI.5.C might have written $a_j a_s a_j^{-1} a_r^{-1} = 1$ and another might have written $a_s a_j^{-1} a_r^{-1} a_j = 1$. However, this does not matter since

$$a_s a_j^{-1} a_r^{-1} a_j = a_j^{-1} a_j a_s a_j^{-1} a_r^{-1} a_j = a_j^{-1} a_j = 1.$$

To cancel out $a_s a_j^{-1} a_r^{-1} a_j$ one might first insert $a_j^{-1} a_j$ before a_s, then cancel out the middle four letters and then the remaining two. We do not need to choose any particular r_i while looking at Figures VI.5.B or VI.5.C. In fact, frequently, instead of reading a relation like $a_j a_s a_j^{-1} a_r^{-1} = 1$ from Figure VI.5.C, it might be simpler to use either $a_r = a_j a_s a_j^{-1}$ or $a_j a_s = a_r a_j$.

EXERCISE VI.5. Suppose $W(f_1) = a_1 a_3^{-1}$, $r_1 = a_1$ is an end point relation, and $r_2 = a_8 a_{10} a_7^{-1} a_{10}^{-1}$ is an undercrossing relation as in Figure VI.3.C. Describe an f_3 so that $f_3 \sim f_1$ and $W(f_3) = a_8 a_{10} a_7^{-1} a_{10}^{-1} a_3^{-1}$.

Solution. Slip the graph of f_1 from under a_1 to get an f_2 such that $f_1 \sim f_2$ and $W(f_2) = a_3^{-1}$. Next, slide the first part of the graph of f_2 so that it comes near the subscript 8 of a_8 in Figure VI.3.C. Push the curve down here and let it cross first under a_8 and a_{10} in direction of the arrows and then under a_7 and a_{10} in directions against arrows, and finally back up near the point of original descent and let if continue as before. There results an f_3 so that $f_2 \sim f_3$ and $W(f_3) = a_8 a_{10} a_7^{-1} a_{10}^{-1} a_3^{-1}$. $\square$

VI.6. Shelling. In this section we digress from considering fundamental groups to look at shelling. Suppose T is a triangulation of a PL disk D^2. A *shelling* of D^2 is an ordering of the 2-simplexes $\Delta_1^2, \Delta_2^2, \ldots, \Delta_n^2$ of T such that if $1 \leqslant n_0 \leqslant n$, $\bigcup_{i=1}^{n_0} \Delta_i^2$ is a PL disk. We say that Δ_n^2 was shelled first and Δ_1^2 saved until last. This last Δ_1^2 is not shelled at all but left instead.

THEOREM VI.6.A. *Any triangulation of a 2-disk can be shelled so that any designated 2-simplex can be saved until last.*

PROOF. The proof is by induction on the number of 2-simplexes in the triangulation T of a disk D^2. We are given Δ_1^2 and it is saved until last.

If some 1-simplex Δ^1 of T spans D^2, it separates D^2 into two disks D_1^2, D_2^2 so that $\Delta_1^2 \subset D_1$ and $D_1 \cap D_2 = \Delta^1$. Then shell D_2 saving until last the 2-simplex of T in D^2 containing Δ^1.

If no 1-simplex of T spans D^2, start by using for Δ_n^2 any 2-simplex in T other than Δ_1^2 with an edge on Bd D^2. $\square$

Instead of triangulating a disk we may subdivide it into subdisks. It is to be understood that if two of the subdisks intersect, the intersection is a subset of the boundary of each. Such a subdivision is called a *cellular subdivision of a disk*. A *shelling* of this subdivision is an ordering $D_1, D_2, \ldots, D_n$ of the subdisks such that each $D_1 \cup D_2 \cup \cdots \cup D_i$ $(1 \leqslant i \leqslant n)$ is a disk. The following is a variation of Theorem VI.6.

THEOREM VI.6.B. *Any cellular subdivision of a 2-disk can be shelled so that any designated disk of the subdivision is saved until last.* $\square$

Shelling off 3-simplexes in dimension 3 plays an important role in the study of 3-dimensional topology, but we postpone a discussion of this until the sixth section of Chapter XIV.

VI.7. Changing words. In this section we show that if $f_1 \sim f_2$ in $E^3 - G$ then $W(f_1)$ can be changed to $W(f_2)$ with a finite number of operations of the following four types:

(1) Inserting xx^{-1} or $x^{-1}x$ anywhere.

(2) Canceling xx^{-1} or $x^{-1}x$ anywhere.

(3) Inserting any r_i or its inverse anywhere.

(4) Canceling any r_i or its inverse anywhere.

It simplifies matters to regard things as polyhedral so we suppose that G is a finite graph, P is a horizontal plane in regular position with respect to G, ρ is the vertical projection into P, $G' = \rho(G)$, and f_1, f_2 are piecewise linear loops normal with respect to G such that $f_1 \sim f_2$ in $R^3 - G$. No vertex of the graph of f_i is on $\rho^{-1}(G')$, and no point of the graph of f_i intersects any vertical line through either a vertex of G or an undercrossing point of G.

Let S be a square disk in R^2 with opposite vertieces $(0,0)$ and $(1,1)$, h_1 be a PL homeomorphism of $[0,1]$ onto the union of the left side and top of S, and g_2 be a PL homeomorphism of $[0,1]$ into the union of the bottom and right side of S. Let F be a homeomorphism of Bd S into R^3 so that $F(h_1) = f_1$ and $F(g_2) = f_2$.

Since $f_1 \sim f_2$ in $R^3 - G$, there is a map $F': S \to R^3 - G$ such that $F' = F$ on Bd S. Let T be a triangulation of S of small mesh and F be a linear map of S under T into $R^3 - G$. At first we let $F = F'$ on vertices on T but then make needed shifts so as to get a linear map $F: S \to R^3 - G$ so that

$$F = F' \quad \text{on Bd } S,$$

and for each 2-simplex Δ^2 of T, if $\rho F(\Delta^2)$ intersects G', it intersects as shown in Figures VI.7.B, VI.7.C, VI.7.D or VI.7.E.

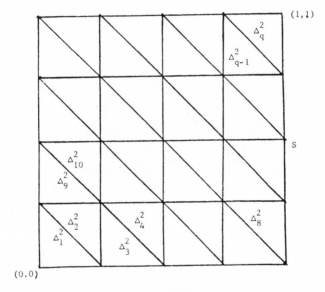

FIGURE VI.7.A

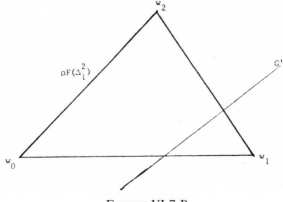

FIGURE VI.7.B

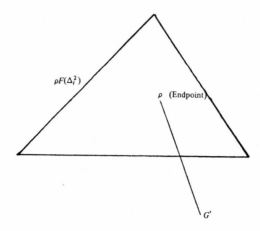

FIGURE VI.7.C

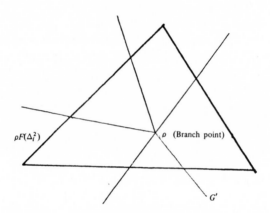

FIGURE VI.7.D

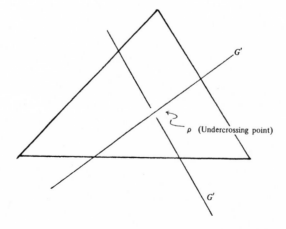

FIGURE VI.7.E

While we could cope with any shelling of any arbitrary triangulation of S, there is some simplicity in triangulating S as shown in Figure VI.7.A and shelling off $\Delta_q^2, \Delta_{q-1}^2, \ldots, \Delta_1^2$ in the order shown. Recall that h_1 took $[0, 1]$ onto the left side and top of S. We suppose Δ_q^2 is shelled and we get a new h (say h_2) that differs from h_1 in that $h_2[0, 1]$ contains the slanted and vertical side of Δ_q^2 rather than the horizontal side. Also, $h_3(0, 1)$ contains vertical and horizontal sides of Δ_{q-1}^2 rather than a slanted side. As i increases, $h_i[0, 1]$ moves down until $h_{q+1}[0, 1]$ is the union of the bottom and right side of S.

Note that $W(f_1) = W(F(h_1))$ and $W(f_2) = W(F(h_{q+1}))$. We complete verification that $W(f_1)$ can be changed to $W(f_2)$ by allowable operations by showing that for each i, $W(F(h_i))$ can be changed to $W(F(h_{i+1}))$ by such operations. We pass from h_i to h_{i+1} by shelling Δ_{q+1-i}.

If $\rho F(\Delta_{q+1-i}^2) \cap G' = 0$, $W(F(h_i)) = W(F(h_{i+1}))$ and no operation in necessary.

If $\rho F(\Delta_{q+1-i}^2) \cap G'$ resembles Figure VI.7.B, $F(\Delta_{q+i-1}^2) \cap G = \varnothing$ so G either passes below or above the triangle since $F(S) \subset R^3 - G$. If it passes below, $W(F(h_i)) = W(F(h_{i+1}))$, and if it passes below, the operation used depends on which sides of Δ_{q+1-i}^2 belong to $h_i[0, 1]$ and $h_{i+1}[0, 1]$. If one of the sides crossing under G belongs to $h_i[0, 1]$ and the other to $h_{i+1}[0, 1]$, $W(F(h_i)) = W(F(h_{i+1}))$ with no change. If both such sides belong to the same one of $h_i[0, 1]$, $h_{i+1}[0, 1]$, an operation of Type 1 or Type 2 changes $WF(h_i)$ to $W(F(h_{i+1}))$.

If $\rho F(\Delta_{q+i-1}^2) \cap G'$ resembles Figure VI.7.C, G misses $F(\Delta_{q+i-1}^2)$. If G is below, $W(F(h_i)) = W(F(h_{i+1}))$, and if it is above, an end point relation can be used to change $W(F(h_i))$ to $W(F(h_{i+1}))$.

Also, if Figure VI.7.D shows the situation, G misses $F(\Delta_{q+i-1}^2)$ and $W(F(h_i)) = W(F(h_{i+1}))$ or can be changed to it by using a branch point relation in conjunction with some operation of Types 1 and 2 to make the branch point operation have the proper start.

If Figure VI.7.E describes the situation, there are more possibilities. If $F(\Delta_{q+i-1}^2)$ is above G, $W(F(h_i)) = W(F(h_{i+1}))$. If it is below, the procedure is the same as in consideration of Figure VI.7.D except that we use an undercrossing relation rather than a branch point relation. If one part of G is beneath $F(\Delta_{q+i-1}^2)$ and another part is above, we ignore the part below and proceed as in the treatment of the case shown by Figure VI.7.B.

Suppose G is a PL finite graph in R^3 and f_1, f_2 are loops in $R^3 - G$ with a common base point. The principle results of §§VI.5 and VI.7 can be stated as follows.

THEOREM VI.7. $f_1 \sim f_2$ in $R^3 - G$ if and only if $W(f_1)$ can be changed to $W(f_2)$ by a finite number of operations of the prescribed types. $\square$

VI.8. Presentation of groups. We use $H = \{a_1, a_2, \ldots\}$ to denote the free group with generators $a_1, a_2, \ldots$. An element of H is an equivalence class of words (finite ordered set of letters) where the letters of the words are the generators $a_1, a_2, \ldots$, and their inverses $a_1^{-1}, a_2^{-1}, \ldots$. Two words belong to the same equivalence class if and only if one can be changed to the other with a finite number of operations where each relation consists of either the insertion of two adjacent letters xx^{-1} or $x^{-1}x$ somewhere in the word, or the cancellation of such a two-letter combination. The identity element is the class which contains the word with no letters.

Suppose $H' = \{a_1, a_2, \ldots \mid r_1 = r_2 = \cdots = 1\}$ is a group, and two words belong to the same equivalence class if one can be changed to the other with a finite number of operations of the following types:

(1) Inserting xx^{-1} or $x^{-1}x$ somewhere.

(2) Canceling xx^{-1} or $x^{-1}x$ somewhere.

(3) Inserting an r_i or its inverse somewhere.

(4) Canceling an r_i or its inverse somewhere.

An alternative way to look at $H' = \{a, a, \ldots \mid r_1, r_2, \ldots = 1\}$ is to let $H' = H/K$ where K is the smallest normal subgroup of H which has elements containing the r_i's.

We find from §§VI.5 and VI.7 that $\pi_1(E^3 - G) = \{a_1, a_2, \ldots, a_n \mid r_1 = r_2 = \cdots = r_m = 1\}$ where the a_i's are the letters attached to the arrows going under arcs in G and the r_i's are the relations obtained at the end points, branch points, and undercrossing points of G.

If G_1, G_2 are two finite graphs in E^3 such that there is a homeomorphism h of E^3 onto E^3 such that $h(G_1) = G_2$, then $\pi_1(E^3 - G_1) = \pi_1(E^3 - G_2)$. However, one should not conclude the converse that if $\pi_1(E^3 - G_1) = \pi_1(E^3 - G_2)$ then there is such a homeomorphism h. Leaving a dangling feeler off a graph does not alter the fundamental group of its complement. In fact, there are smooth simple closed curves (tame knots) in E^3 such that the fundamental groups of their complements are the same even though there is no space homeomorphism taking one onto the other. However, the difference of the fundamental groups of their complements does imply the lack of a space homeomorphism.

There are rules for changing the presentation of a group but we shall not dwell on them here. We can add a generator if we add a relation to tell what the generator equals; a relation may be added that is a consequence of other relations; a relation may be deleted that is a consequence of other relations; if a relation tells what a generator equals, then the generator and the relation may be deleted if no other relation involves the generator.

VI.9. Short cuts. We mention several simplifictions that one discovers as he deals with the fundamental group of the complement of a graph in E^3. The first three involve changing the graph and the last three involve simplifying the presentation of the group.

(a) *Simplifying projections.* If the regular projection of a graph G_1 looked like G_1' as shown in the left half of Figure VI.9.a its complement would have the same fundamental group as the complement of graph G_2 obtained from G_1 by a homeomorpism of E^3 onto itself. The projection G_2' of G_2 shown in the right half of Figure VI.9.a is simpler.

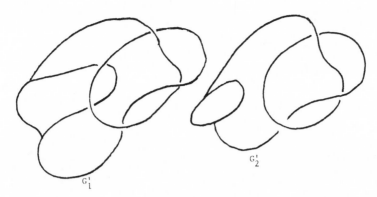

FIGURE VI.9.a

(b) *Shrinking arcs.* If a smooth arc in E^3 is shrunk to a point, the resulting decomposition space is topologically E^3. Let G_2 be a finite graph associated with the projection G_2' shown on the right side of Figure VI.9.a. There is an arc in G_2 such that if this arc is shrunk to a point, G_2 becomes a finite graph G_3 whose projection is shown in the left side of Figure VI.9.b. Note that $E^3 - G_2$ is topologically equivalent to $E^3 - G_3$ and $\pi_1(E^3 - G_2) = \pi_1(E^3 - G_3)$.

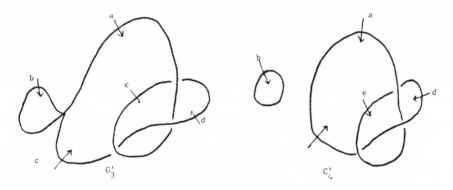

FIGURE VI.9.b

Shrinking arcs also enables us to eliminate dangling arcs from finite graphs. See Figure VI.5.A.

(c) *Pulling graphs apart.* Note that the relation $bb^{-1}ca^{-1} = 1$ obtained at the branch point of G_3 merely tells us that $a = c$. Since no other relation involves b, $\pi_1(E^3 - G_3) = \pi_1(E^3 - G_4)$, where the projection G_4' of G_4 is shown in the right half of Figure VI.9.b. In general, it is permissible to pull a graph G apart at a vertex v if the projection of v is a separating point of the regular projection G' of G. To compute $\pi_1(E^3 - G_1)$, it would be a short cut to compute $\pi_1(E^3 - G_4)$. One should not conclude that since $\pi_1(E^3 - G_3) = \pi_1(E^3 - G_4)$ that $E^3 - G_3$ and $E^3 - G_4$ are homeomorphic.

(d) *Dropping a relation.* Instead of pulling a path f under a particular branch point or an undercrossing point as suggested in Figures VI.5.B or VI.5.C, it is possible to get an equivalent homotopy by pulling the path f to the path f' by moving under the rest of G. Hence any particular branch point or undercrossing relation r_i of $\pi_1(E^3 - G) = \{a_1, a_2, \ldots, a_n \mid r_1 = r_2 = \cdots = r_m = 1\}$ is a consequence of the others and there is no loss in dropping it from the presentation $\{a_1, a_2, \ldots, a_n \mid r_1 = r_2 = \cdots = r_m = 1\}$.

(e) *Equating generators.* Suppose G is a graph whose regular projection G' is shown in Figure VI.9.e. If one followed the suggestions made earlier, one would assign different letters to the vectors a_i as shown. However, one finds by either moving under the part of G on the left, or on the right, that a path going under the lower arc is homotopic to one going under the upper arc. If one writes down the presentation of $\pi_1(E^3 - G)$ as suggested in the preceding section, it follows from the algebra that $a_1 = a_2$. It is a short cut to designate each of a_1, a_2 by the same letter, say a, without going through the algebra.

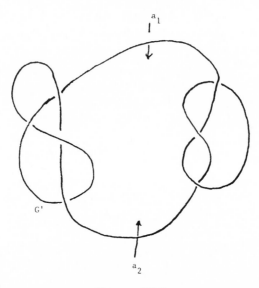

a_1

a_2

G'

FIGURE VI.9.e

(f) *Further elimination of generators and relations.* Consider a graph G whose projection G' is shown in Figure VI.9.f. Instead of giving the vector c a name, it would be a short cut to call it $b^{-1}ab$. It is convenient not to assign separate letters to vectors associated with arcs of G that do not pass above points of G, since these can be assigned words instead of letters. In Figure VI.9.f we could drop c as a generator and combine the relations $c = b^{-1}ab$ and $c = d^{-1}ed$ to become the single relation $b^{-1}ab = d^{-1}ed$.

As a consequence of this observation, one obtains the following.

EXERCISE VI.9.A. Suppose G is a simple closed curve which is the union of n vertical spanning segments of a cube C and n arcs on the boundary of C. Show that $\pi_1(E^3 - G)$ has a presentation with n generators and $n - 1$ relations.

Solution. In Figure VI.9.A we show the cube C with the front face smaller than the back and the curve G missing the back face. The generators associated with the arcs on the boundary of C are the only ones needed since others do not pass in front of other points of G. Hence $\pi_1(R^3 - G)$ can be written with n generators (and inverses). If one starts at the top of vertical segments and assigns words for each arc, there result n-relations. However, one relation is unnecessary. □

EXERCISE VI.9.B. Suppose G is smooth simple closed curve with n undercrossing points. If $n \geqslant 2$, then $\pi_1(E^3 - G)$ has a presentation with $n - 1$ generators and $n - 2$ relations.

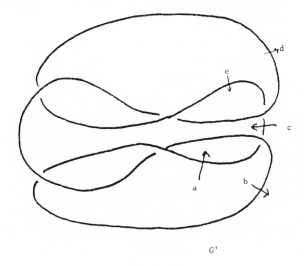

FIGURE VI.9.f

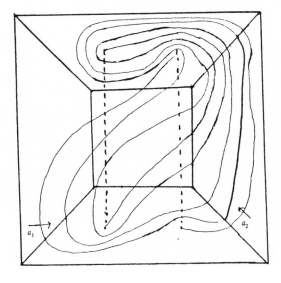

FIGURE VI.9.A

Solution. The n undercrossing points divide G into n arcs $A_1, A_2, \ldots, A_n$. Let $a_1, a_2, \ldots, a_n$ be generators associated with the A_i's.

If some A_i does not pass above an undercrossing point, the generator a_i can be written as the product of other a's, and the two relations involving a_i can be changed to a single relation not involving a_i. Also, we can ignore another relation so to end up using only $n-1$ generators and $n-2$ relations.

If each A_i passes above an undercrossing point, no one passes above two. If A_j passes above $\overline{A}_1 \cap \overline{A}_2$, we can replace a_1 by the appropriate three-letter word and drop the corresponding relation. Dropping another relation gets us down to $n-1$ generators and $n-2$ relations. □

EXERCISE VI.9.C. Can the polygon P shown in Figure VI.9.C be shrunk to a point in the complement of the simple closed curve J shown there?

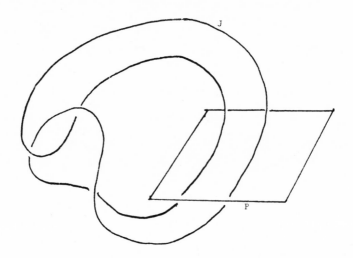

FIGURE VI.9.C

Answer. Yes. A computation of $\pi_1(R^3 - J)$ shows that it is the infinite cyclic group and P corresponds to the identity element. □

VI.10. Why compute fundamental groups? Some mathematicians study the fundamental groups of the complements of curves for their own sake and because they find the study interesting. We point out some applications of the study to the topology of 3-manifolds.

To show that the exterior of the Alexander horned sphere is not simply-connected, one might show that the simple closed curve J of Figure VI.10.A cannot be shrunk to a point in $R^3 - X$.

EXERCISE VI.10.A. Compute and simplify $\pi_1(R^3 - X)$ of Figure VI.10.A.

Solution. $\pi_1(R^3 - X) = \{a, b, c, d, e \mid ecd^{-1} = eab^{-1} = bda^{-1}d^{-1} = ada^{-1}c^{-1} = 1\}$. Throw out e, b, c and the last relation to get $\pi_1(R^3 - X) = \{a, d\}$. Hence $\pi_1(R^3 - X)$ is a free group on 2-generators and J corresponds to the word $ada^{-1}d^{-1}$. □

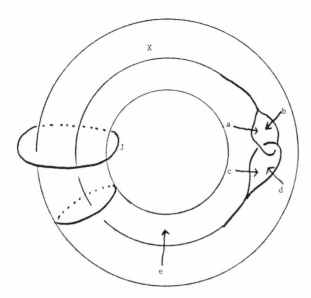

FIGURE VI.10.A

Sometimes the presentation for π_1 is so complicated that it is difficult to determine if the group is trivial. Even if $\pi_1(R^3 - G)$ is nontrivial, there is difficulty in deciding which loops in it are homotopic to constants. The loop theorem from Chapter XVII sometimes offers help. The proof of the following theorem illustrates this.

THEOREM VI.10.B. *Suppose C is the cylinder shown in Figure* VI.10.B *with bases* B_1, B_2; T_1, T_2 *are two eye bolts embedded in C as shown so that* T_1 *and* T_2 *link while* $T_i \cap \mathrm{Bd}\, C = B_i$; *and X is a closed set in* R^3 *such that* $X \cap C = B_1 \cup B_2$. *If J is a simple closed curve in* $R^3 - (X \cup C)$ *that cannot be shrunk to a point in* $R^3 - (X \cup C)$, *then J cannot be shrunk to a point in* $R^3 - (X \cup T_1 \cup T_2)$.

PROOF. Assume that J can be shrunk to a point in $R^3 - (X \cup T_1 \cup T_2)$. Suppose J and Bd C are PL. Let T be a triangulation of Δ^2 and f be a linear map of (Δ^2, T) into $R^3 - (X \cup T_1 \cup T_2)$ such that f takes Bd Δ^2 homeomorphically onto J, f is a homeomorphism on each simplex Δ of T, and each $f(\Delta)$ is GP with respect to Bd C. Each component of $f^{-1}(\mathrm{Bd}\, C)$ is a simple closed curve on Δ^2. Induction on the number of components of $f^{-1}(\mathrm{Bd}\, C)$ is used to show that f does not exist. Assume $f^{-1}(\mathrm{Bd}\, C)$ has k components and Theorem VI.10.B is true if it has fewer. Let J_1 be a simple closed curve of $f^{-1}(\mathrm{Bd}\, C)$ that bounds an innermost disk D_1 in Δ^2. Then f on J_1 is not homotopic to a point on Bd $C - (B_1 \cup B_2)$; otherwise we could redefine f near D_1 and eliminate a component of $f^{-1}(\mathrm{Bd}\, C)$.

Although J_1 is a simple closed curve, $f(J_1)$ may be singular. However, it follows from the loop theorem (see Chapter XVII) that there is a simple closed curve K on Bd $C - (B_1 \cup B_2)$ such that K does not bound a disk on Bd $C - (B_1 \cup B_2)$ but K does bound a disk E in the one of $C - (T_1 \cup T_2)$ or $R^3 - (X \cup \mathrm{Int}\, C)$ containing $f(D_1)$. We show that this is impossible.

The simple closed K on Bd C separates B_1 from B_2 since it does not bound a disk on Bd $C - (B_1 \cup B_2)$. It follows from Exercise VI.10.A that $E \not\subset C - (T_1 \cup T_2)$.

If $E \subset R^3 - (X \cup \text{Int } C)$, then J can be shrunk to a point in $R^3 - (X \cup \text{Int } C)$. The assumption that there was an f led to the contradiction that there is no place for E. $\square$

THEOREM VI.10.C. *The complement of the solid Alexander horned sphere is not simply-connected.*

PROOF. Exercise VI.10.A shows that the simple closed curve P of Figure IV.3.C cannot be shrunk to a point in $R^3 - H_1$. It follows from repeated applications of Theorem VI.10.B that is cannot be shrunk to a point in any $R^3 - H_i$. The compactness of the image of Δ^2 under a map shows that it cannot be shrunk to a point in the complement of $R^3 - \cap H_i$. $\square$

Suppose X of Figure VI.10.D is a modification of the Fox-Artin arc as shown in Figure IV.9.A where a solid cone C_0 to the left of F_1 and a solid cone C_1 to the right of F_1 are added. To help compute $\pi_1(R^3 - X)$ we have substituted dotted triods for the cones.

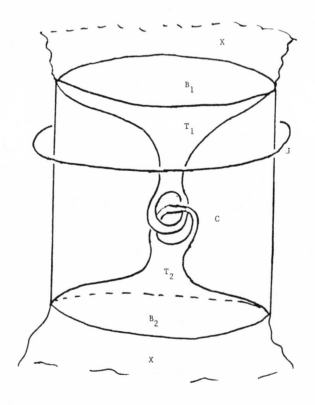

FIGURE VI.10.B

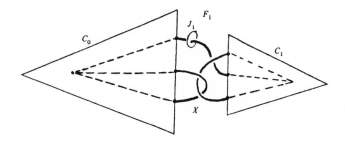

FIGURE VI.10.D

EXERCISE VI.10.D. Compute $\pi_1(R^3 - X)$.

Answer. $\pi_1(R^3 - X)$ is a free group on two generators.   □

The following result follows from an algebraic computation of $\pi_1(R^3 - X)$.

EXERCISE VI.10.E. The simple closed curve J_1 shown in Figure VI.10.D cannot be shrunk to a point in $R^3 - X$.   □

We are now ready to prove that the complement of the Fox-Artin arc of Figure 9.A is not simply-connected. Let A denote the arc and C_0, C_1, C_2 be the cones shown in Figure VI.10.F.

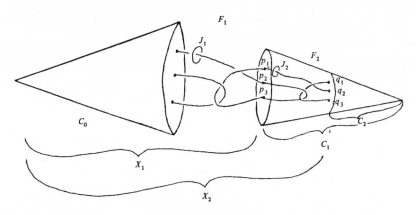

FIGURE VI.10.F

EXERCISE VI.10.F. J_1 cannot be shrunk to a point in $R^3 - (A \cup C_0 \cup C_2)$.

PROOF. Following the proof of Theorem VI.10.B, we find that the assumption that J_1 can be shrunk to a point in $R^3 - (A \cup C_0 \cup C_2)$ leads to a disk E in $R^3 - (A \cup C_0 \cup C_2)$ such that $E \cap \text{Bd } C_1 = \text{Bd } E$ and Bd E does not bound a disk on Bd $C_1 - A = \text{Bd } C_1 - \{p_1, p_2, p_3\}$. Again we find that E cannot be in C_1 or in $R^3 - (A \cup C_0 \cup \text{Int } C_1)$.

We proved Exercise VI.10.F by showing there is no place for E.   □

The following result can be proved from repeated application of Exercise VI.10.F.

THEOREM VI.10.G. *The complement of the Fox-Artin arc shown in Figure IV.9.A is not simply-connected.*   □

The following two exercises are related to Antoine's necklace.

EXERCISE VI.10.H. The simple closed curve J shown in Figure VI.10.H cannot be shrunk to a point in the complement of $T_1 \cup T_2$ shown there.

Solution. $\pi_1(R^3 - (T_1 \cup T_2)) = \{a, b\}$ and J corresponds to the word $bab^{-1}a^{-1}$. $\square$

EXERCISE VI.10.I. If there are three or more T_i's in Figure VI.10.I, then J cannot be shrunk to a point in $R^3 - \cup T_i$.

PROOF. Let M_1 be a plug with two holes filling T_1 as shown. First we show that if J can be shrunk to a point in $R^3 - (\cup T_i)$, it can be shrunk in $(R^3 - \cup T_i) - M_1$.

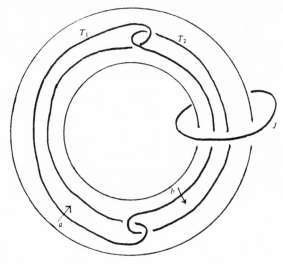

FIGURE VI.10.H

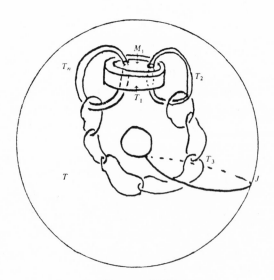

FIGURE VI.10.I

If J can be shrunk in $R^3 - \cup T_i$ but not in $(R^3 - \cup T_i) - M_1$, it follows from the argument used in the proof of Theorem VI.10.B that there is a disk E in $M_1 - \cup T_i$ such that $E \cap \text{Bd } M_1 = \text{Bd } E \subset \text{Bd } M_1 - (\cup T_i)$, and Bd E does not bound on Bd $M_1 - (\cup T_i)$. Then one or both of T_2, T_n prevent D_i from lying either in M_1 or $R^3 - M_1$. Hence J can be shrunk to a point $(R^3 - \cup T_i) - M_1$.

Let M_2 be a plug with one hole which finishes filling T_2. An argument similar to that used above shows if J can be shrunk in $R^3 - \cup T_i$ it can be shrunk in $(R^3 - \cup T_i) - (M_1 \cup M_2)$. By putting in more plugs we find that if J can be shrunk to a point in $R^3 - \cup T_i$, it can be shrunk to a point in the complement of a center line of the big torus T. This, of course, is impossible. $\square$

EXERCISE VI.10.J. Can the polygon P of Figure VI.10.J be shrunk to a point in $R^3 - (T_{11} \cup T_{12})$?

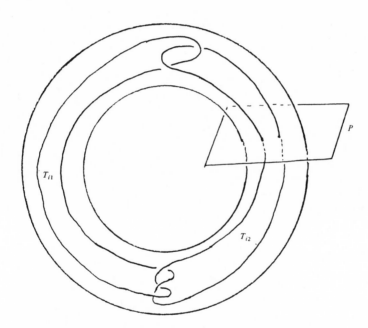

FIGURE VI.10.J

Answer. Yes. A computation of $\pi_1(R^3 - (T_{11} \cup T_{12}))$ shows that it is a free abelian group on 2-generators and P corresponds to the identity element.

It is an unsolved question (the *Poincaré conjecture*) as to whether or not each compact connected 3-manifold whose fundamental group is trivial is a topological 3-sphere. In trying to construct counterexamples, one might remove from S^3 a tubular neighborhood of a smooth finite graph in S^3 and sew this tubular neighborhood back in differently. It is known that if the Poincaré conjecture is false, a counterexample can be obtained in this way. The fundamental group of the complement of the tubular neighborhood is the same as the fundamental group of the complement of the graph. It is not known if the study of fundamental groups will be of any help in solving the Poincaré conjecture, but hopes that it may have encouraged an interest in these groups.

There is a distinction between being able to shrink a curve to a point and the curve bounding a disk. Algebra is sometimes of help in deciding whether or not a map is homotopic to a constant. For example, it shows that P of Figure VI.9.C shrinks in $R^3 - J$, but it does not tell us if it bounds a disk there. It does not tell us if P bounds a singular disk in $R^3 - J$ whose set of singularities is topologically starlike. We seem to need stronger tools to decide the Poincaré conjecture. Perhaps there is a 3-manifold whose fundamental group is presented in some complicated exotic fashion and that by some clever trick the group can be shown to be trivial. If the clever trick did not cause simple closed curves to lie in 3-cells in the 3-manifold, we would have a counterexample to the Poincaré conjecture.

VI.11. A homology cube. The cube with a knotted hole shows in Figure VI.11 is denoted by M^3. A center line for the knotted hole is provided with short oriented normal vectors a, b, c to facilitate the computation of $\pi_1(M^3)$ by methods we have used earlier in this chapter. One finds that $\pi_1(M^3) = \{a, b, c \mid aba^{-1}c^{-1} = bca^{-1}c^{-1} = 1\}$. While this presentation can be simplified, it is not our purpose to do so.

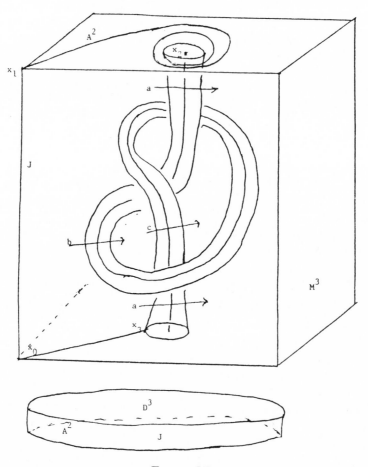

FIGURE VI.11

Let J be the simple closed curve on Bd M^3 associated with the loop which rises from x_0 to x_1, goes from x_1 to x_2 on the top while circling the hole twice as shown, goes down the side of the hole to x_3, and returns to x_0 on the bottom. Then J corresponds to the word $a^2c^{-1}a^{-1}b^{-1}$.

Let P^3 be the pill box $D^2 \times I$ shown in the lower part of Figure VI.11, where its lateral side is denoted by A_2^2. §XVII.1 has a definition of a pillbox. Use A_1^2 to denote an annulus on Bd M^3 whose center line is J and let h be a homeomorphism of A_1^2 onto A_2^2. Suppose M^3 is sewed to P^3 along A_1^2 and A_2^2 by h as described in IV.2.B and IV.2.D. Then $K^3 = M^3 \cup_h P^3$ has a 2-sphere as a boundary and $\pi_1(K^3) = G = \{a, b, c \mid aba^{-1}c^{-1} = bca^{-1}c^{-1} = a^2c^{-1}a^{-1}b^{-1}\}$. One may note that if G is abelianized by adding relations $ab = ba$, $ac = ca$, $bc = cb$, then the first two relations imply $a = b = c$ and the last implies $a = 1$. Hence, the abelianization of $\pi_1(K_3)$ is trivial.

Sometimes K^3 is called a *homology cube*. One could get many other examples of such objects by boring different holes in cubes, but then x_1x_2 might need to circle the hole a different number of times to make the abelianization of the fundamental group trivial. Some require that the homology cube be topologically different from a 3-cell but we do not.

Question. Is there some nontrivial knotted hole in a cube such that if a pill box is added as described, then the fundamental group of the union is trivial?

VI.12. Other treatments. The techniques we have used for computing the fundamental group of the complement of graphs might be called the shelling method after our treatment of Theorem VI.7. This treatment [**B₂₃**] is covered in a Technical Report issued by Washington State University in 1965 by R. H. Bing entitled "Computing the Fundamental Group of the Complement of Curves". There is another widely used method where we enlarge G by digging more tunnels so that the fundamental group of the complement of the enlarged graph is free abelian. The extra tubes are then plugged and a theorem by van Kampen is used to compute $\pi_1(R^3 - G)$. Treatments of this technique are given by Crowell and Fox [**CF**], Fox [**F₃**], and Rolfsen [**R₁**].

CHAPTER VII

MAPPING ONTO SPHERES

Mathematicians have used many approaches to discover the topological properties of objects in Euclidean spaces. In Chapter VI we considered the mappings of circles. In this chapter we consider a converse operation of mapping sets onto spheres. In their excellent book, *Dimension Theory*, Hurewitz and Wallman verified many topological properties of Euclidean sets by mapping them onto spheres. Chapter VI of [**HW**] is especially pertinent.

Whether or not each map of a compact subset X of R^2 onto the unit circle S^1 extends to a map of R^2 onto S^1 gives us important information about whether or not X separates R^2. What we learn provides a proof of the Jordan curve theorem, and an extension of the method gives the n-dimensional version.

Recall that S^{n-1} is *the canonical unit $(n-1)$-sphere* in R^n centered at the origin. It has vector equation $|x| = 1$.

All sets X, Y, Z in this chapter are assumed to lie either in some *Euclidean space* R^n or *Hilbert space* R^∞. We use R^m to denote either.

VII.1. Retractions onto boundaries. If X is a subset of Y, a map f of Y onto X is called a *retraction* if for each $x \in X$, $f(x) = x$.

THEOREM VIII.1.A. *There is no retraction of a 2-simplex onto its boundary.*

PROOF. We show first that if there is a retraction $r : \Delta^2 \to \mathrm{Bd}\ \Delta^2$, then there is a triangulation T of Δ^2 and a linear retraction of (Δ^2, T) onto $\mathrm{Bd}\ \Delta^2$ such that the image of each simplex of T lies on a face of Δ^2. We obtain T by triangulating Δ^2 so that no 1-simplex of T spans Δ^2 and the mesh of T is very small.

We define f as follows. If x is a point of $\mathrm{Bd}\ \Delta^2$, $f(x) = x$; if v is an interior vertex of T, $f(v)$ is the nearest end of the 1-face of Δ^2 containing $r(v)$ if there is a nearest end; otherwise, use either end; f is linear on each simplex of T.

Finally, we show that this linear retraction cannot exist. Let $\rho \in \mathrm{Bd}\ \Delta^2 - f(T^0)$ where T^0 is the 0-skeleton of T. Consider $f^{-1}(p)$. It starts out as an arc from p but has no other end. Every compact 1-manifold has an even number of ends, but the assumption that there is a retraction of Δ^2 onto Bd Δ^2 led to the contradiction that there is one with only one end.   □

The preceding theorem generalizes to the following

THEOREM VII.1.B. *There is no retraction of a compact triangulated n-manifold with boundary onto its boundary.*   □

VII.2. AR's and ANR's. A compact set X is an *absolute retract* (AR) if for each homeomorphism h of X into a set Y, there is a retraction r of Y onto $h(X)$. An *n-cell* I^n is the Cartesian product of n 1-simplexes and is homeomorphic to Δ^n.

THEOREM VII.2.A. I^n *is an AR.*

PROOF. Suppose Y lies on R^m. It follows from the Tietze's extension theorem that there is a map g of R^m into I^n that agrees with h^{-1} on $h(I^n)$. Then hg restricted to Y is a retraction of Y to $h(I^n)$.   □

THEOREM VII.2.B. *Suppose X is an AR which lies in Y and f is a map of X into Z. Then f can be extended to map of Y to Z.*

PROOF. A suitable map is fr where r is a retraction of Y onto X.   □

A compact set X is called an *absolute neighborhood retract* (ANR) if for each imbedding h of X into Y there is an open set U in Y containing $h(X)$ such that U retracts onto $h(X)$.

THEOREM VII.2.C. S^{n-1} *is an ANR.*

PROOF. Suppose $h(S^{n-1})$ lies in Y which in turn lies in R^m. The Tietze's extension theorem implies that the map $h^{-1}: h(S^{n-1}) \to R^n$ can be extended to a map $g: R^m \to R^n$. Let π be the projection from the origin p_0 of R^n sending $R^n - p_0$ onto S^{n-1}. Then $h\pi g$ takes $R^m - g^{-1}(p_0)$ into $h(S^{n-1})$ and hence takes an open subset of Y into $h(S^{n-1})$.   □

THEOREM VII.2.D. *Suppose X is an ANR, $X \subset Y$, and f is a map of X into Z. Then f can be extended to take an open subset U of Y onto Z.*   □

THEOREM VII.2.E. *Suppose X is a compact subset of Y and f is a map of X into Z where Z is an AR. Then f can be extended to take Y into Z.*

PROOF. We suppose $Y \subset R^m$ and $Z \subset R^k$. Tietze's extension theorem provides a map g of R^m into R^k that extends f. If r is the retraction of R^k onto Z, a suitable map is rg restricted to Y.   □

THEOREM VII.2.F. *Suppose X is a compact subset of Y and f is a map of X into Z where Z is an ANR. Then there is an open set U in Y containing X such that f extends to take U to Z.*   □

VII.3. Extending mappings onto spheres. Suppose $X \subset Y$ and f is the map of X onto a sphere S^{n-1}. Can f be extended to map Y into S^{n-1}? Answers to such questions enables one to learn whether or not sets in R^n separate it.

A PL *neighborhood* of an object X in R^n is an open set in R^n containing X which is the union of a finite number of vertices and interiors of simplexes of some triangulation T of R^n.

THEOREM VII.3.A. *Suppose X is a compact subset of R^n, U is the unbounded component of $R^n - X$, and f is a map of X into S^{n-1}. Then f can be extended to send U into S^{n-1}.*

PROOF. The proof is similar for all dimensions, but for simplicity we treat it for $n = 2$. It follows from Theorems VII.2.C and VII.2.F that X lies in an open set U' of R^2 such that f can be extended to U'. We suppose with no loss of generality that U' lies in a large 2-simplex Δ^2 of R^2 centered at the origin.

Cover the closed set $\Delta^2 \cap (U - U')$ with a PL neighborhood N that misses X. We see that we can make N connected by digging tunnels and dropping unneeded components, so suppose with no loss of generality that N is connected and Bd Δ^2 lies in the 1-skeleton T^1 of the triangulation T used in determining N. Let σ_1^2, $\sigma_2^2, \ldots, \sigma_k^2$ be an ordering of the 2-simplexes of T whose interiors lie in $N \cap \Delta^2$ so that σ_1^2 has a 1-face σ_1^1 on Bd Δ^2, and for $1 \leqslant i < k - 1$, σ_{i+1}^2 has a face σ_{i+1} on Bd $\Delta^2 \cup \sigma_1^2 \cup \cdots \cup \sigma_i^2$.

Recall that we extended f to map U' onto S^1. Let g be this extension restricted to $(X \cup U) \cap (\Delta^2 - N)$. We work at the finite sequence $\sigma_1^2, \sigma_2^2, \ldots, \sigma_k^2$ backwards and extend g to σ_k^2, then to $\sigma_{k-1}^2, \ldots$, and finally to σ_1^2. To extend g to σ_i^2, first extend it to Bd σ_i^2 - Int σ_i^1 (if any extension is necessary), and then apply the previous extension to projection of σ_i^2 onto Bd σ_i^2 - Int σ_i^1. After the final extension to projection of σ_i^2 onto Bd σ_i^2 - Int σ_i^1. After the final extension to σ_1^2 we have sent $(X \cup U) \cap \Delta^2$. To extend it to the part of U outside Δ^2 we project Ext Δ^2 toward the center of Δ^2 onto Bd Δ^2 and then apply the extension. $\square$

THEOREM VII.3.B. *If a closed compact set C in R^n does not separate R^n, any map of C into S^{n-1} can be extended to map R^n into S^{n-1}.* $\square$

THEOREM VII.3.C. *If X is a closed subset of I^{n-1}, any map of X into S^{n-1} extends to take I^{n-1} into S^{n-1}.*

PROOF. Let T be triangulation of $I^{n-1} - X$ such that the diameters of each simplex is less than its distance from X. Suppose the given map of X into S^{n-1} is first extended to $X \cup T^0$ and then a map f of I^{n-1} into R^n that is linear on each simplex of T and agrees with the given map on X. Since $f(I^{n-1} - X)$ lies in union of a countable number of $(n-1)$-planes in R^n, there is a point p_0 in $R^n - f(I^{n-1})$ in Int S^{n-1}. If r is the projection from p_0 onto S^{n-1}, rf is a suitable map. $\square$

VII.4. Inessential mappings. Two mappings f, g of X into Y are *homotopic* if there is a map $F : (X \times [0, 1]) \to Y$ such that $F(x, 0) = f(x)$, $F(x, 1) = g(x)$. We call F a homotopy and frequently think of it as a 1-parameter family of maps $f_t : X \to Y$, $t \in [0, 1]$ where $f_t(x) = F(x, t)$. We say that f is *inessential* if it is homotopic to a constant map g. ($g(X)$ is a 1-point set.) If f is not inessential, it is *essential*.

THEOREM VII.4.A. *Any map of an AR into a space Y is inessential.*

PROOF. Suppose X is an AR in I^m. It follows from Theorem VII.3.B that the given map can be extended to a map f of I^m into Y.

Let $\phi : (I^m \times [0, 1]) \to I^m$ be a shrinking homotopy where $\phi(x \times 0) = x$, and $\phi(I^m \times 1)$ is a 1-point set. Then $f\phi : (X \times [0, 1]) \to Y$ is a suitable homotopy. $\square$

THEOREM VII.4.B. *Any map f of a closed subset X of I^{n-1} into S^{n-1} is inessential.*

PROOF. It follows from Theorem VII.3.C that a map f of X into S^{n-1} can be extended to a map g of I^{n-1} into S^{n-1}. The map g is inessential by Theorem VII.4.A and the homotopy applied to X shows that f is also. $\square$

THEOREM VII.4.C. *If X is a compact subset of R^n which does not separate R^n, then any map of X into S^{n-1} is inessential.*

PROOF. Let I^n be an n-cell in R^n whose interior contains X. It follows from Theorem VII.3.A that a map f of $X \to S^{n-1}$ can be extended to take I^n to S^{n-1}. But I^n is an AR and Theorem VII.4.A promises a homotopy f_t sending the extended f to a constant map. This same homotopy applied to X sends f to a constant map. $\square$

The following theorem is an important one in homotopy theory.

THEOREM VII.4.D (HOMOTOPY EXTENSION THEOREM). *Suppose X is a compact subset of Y, Z is an ANR, and f is an inessential map of X into Z. Then f can be extended to send Y into Z.*

PROOF. Consider $Y \times [0, 1]$ as shown in Figure VII.4.D. Let f' be a map of $X \times [0, 1] \cup (Y \times 1)$ into Z such that $f'(x \times 0) = f$ and $f'(Y \times 1)$ is a point. It follows from Theorem VII.2.F that there is an open set U in $Y \times [0, 1]$ containing $X \times [0, 1] \cup (Y \times 1)$ so that f' can be extended to take U into Z. We call the extension f.

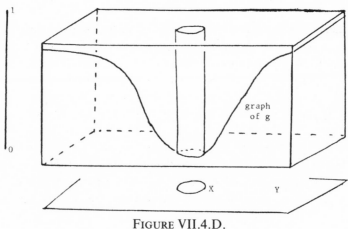

FIGURE VII.4.D.

It is an exercise in general topology to show that there is a map g of Y into $[0, 1]$ such that $g(X) = 0$, and the part of $Y \times [0, 1]$ above the graph of g lies in U. Then the extension of Y to Z is obtained by taking the vertical projection of $Y = Y \times 0$ onto this graph of g and then following this projection with f. $\square$

Figure VII.4.D indicates that the proof of the homotopy extension theorem is suggested by an upside down bell-shaped distribution graph. Similarly, the proof of Borsuk's Theorem VII.4.E is suggested by a noninverted such graph.

THEOREM VII.4.E (BORSUK'S THEOREM). *Suppose X is a closed subset of Y and f_0, f_1 are homotopic maps of X into an ANR Z. If f_0 can be extended to X, then f_1 can be so that the two extensions are homotopic.* $\square$

VII.5. Projections. If p is the origin in R^n, we use π_p to denote the projection from p of $R^n - p$ onto S^{n-1}. Sometimes π_p is called a radial projection from p. If p is any other point of R^n we let π_p act on $R^n - p$ by first projecting $R^n - p$ from p onto the $(n-1)$-sphere $|x - p| = 1$ and then translating this $(n-1)$-sphere to S^{n-1}. As a vector equation we say $\pi_p(x) = (x - p)/|x - p|$.

THEOREM VII.5.A. *Suppose X is a compact subset of R^n and $R^n - X$ has a bounded component U. If $p \in U$, $\pi_p : X \to S^{n-1}$ is essential.*

PROOF. With no loss of generality we suppose p is origin and $U \subset \text{Int } S^{n-1}$.

Assume π_p is inessential. Theorem VII.4.D implies there is an extension $g : R^n \to S^{n-1}$ of $\pi_p : X \to S^{n-1}$. Consider the map $r : S^{n-1} \cup \text{Int } S^{n-1} \to S^{n-1}$ described as follows: $r = g$ except on unbounded component of $R^n - X$; $r = \pi_p$ on the remainder of $S^{n-1} \cup \text{Int } S^{n-1}$. The assumption that $\pi_p : X \to S^{n-1}$ is inessential led to the contradiction to Theorem VII.1.B which assures that there is not a retraction of $S^{n-1} \cup \text{Int } S^{n-1}$ onto S^{n-1}.

VII.6. Separating R^n. Theorem VII.3.A implies the following.

THEOREM VII.6.A. *If a compact subset C of R^n fails to separate R^n, each map of C into S^{n-1} is inessential.* $\square$

The following result follows Theorems VII.6.A and VII.5.A.

THEOREM VII.6.B. *A compact subset C of R^n separates R^n if and only if there is an essential map of C onto S^{n-1}.* $\square$

THEOREM VII.6.C. *A compact C of R^n separates R^n if and only if each of its homeomorphic images in R^n separates.* $\square$

The theorem might be restated as follows.

THEOREM VII.6.C. *The ability to separate R^n is a topological property for compact sets.* $\square$

The following result follows from Theorems VII.3.C, VII.4.A, VII.6.B, or from Theorem VII.6.C.

THEOREM VII.6.D. *If X is the homeomorphic image in R^n of a compact subset of I^{n-1}, X does not separate R^n.* □

THEOREM VII.6.E. (*n-DIMENSIONAL JORDAN CURVE THEOREM*). *Any topological $(n-1)$-sphere Σ^{n-1} in R^n separates R^n and Σ^{n-1} is the boundary of each component of $R^n \doteq \Sigma^n$.*

PROOF. That Σ^{n-1} separates R^n follows from Theorem VI.6.C. That Σ^{n-1} is the boundary of each component of $R^n - \Sigma^{n-1}$ follows from the fact that no topological $(n-1)$-cell separates R^n (Theorem VII.6.D). □

We leave until later the task of showing that $R^n - \Sigma^{n-1}$ has precisely two components. We proved this in Chapter III for the case where $n = 2$, but there we were using special arguments that do not generalize easily to higher dimensions.

THEOREM VII.6.F (*INVARIANCE OF DOMAIN*). *Suppose X is a closed subset of R^n, h is a homeomorphism of X into R^n, and $x \in X$. Then $h(x) \in \mathrm{Bd}\, h(X)$ if and only if $x \in \mathrm{Bd}\, X$.*

PROOF. Assume $x \in \mathrm{Bd}\, X$ and $h(x) \in h(X) - \mathrm{Bd}\, h(X)$. Then in each open n-ball N in R^n containing x there is a closed proper subset Y of an $(n-1)$-sphere such that Y separates two points of $N \cap X$ from each other in X. However, Theorem VII.6.D says that for Y sufficiently small, $h(Y)$ does not separate any two points in $h(X)$. Similarly we find that connectivity causes $h(x) \in \mathrm{Bd}\,(X)$ to imply that $x \in \mathrm{Bd}\, X$. □

The requirement that X was closed was included for convenience rather than necessity. Hence Theorem VII.6.F can be stated as follows.

THEOREM VII.6.G. *If h is a homeomorphism of a subset X of R^n into R^n and $x \in X$, then $x \in \mathrm{Bd}\, X$ if and only if $h(x) \in \mathrm{Bd}\, h(X)$.* □

VII.7. Fixed points. One of the most interesting applications of the result (Theorem VII.1.B) that there is no retraction of an n-cell ($n \geqslant 1$) onto its boundary is the result that each n-cell ($n \geqslant 1$) has the fixed-point property. A set X has the *fixed-point property* if each map $f: X \to X$ has a fixed point—that is, a point $x_0 \in X$ such that $f(x_0) = x_0$. The point x_0 is a function of f.

THEOREM VII.7.A. *Each n-cell has the fixed-point property.*

PROOF. It is convenient to regard the n-cell as $S^{n-1} \cup \mathrm{Int}\, S^{n-1}$ so we do this. If there is a fixed-point free map $f: S^{n-1} \cup \mathrm{Int}\, S^{n-1} \to S^{n-1} \cup \mathrm{Int}\, S^n$ let $r(x)$ be the point where the open ray from $f(x)$ through x hits S^{n-1}. It can be shown that r is continuous. The assumption that there is a fixed-point free map f of $S^n \cup S^{n-1}$ onto itself led to the contradiction that there is a retraction r of an n-cell onto its boundary. □

VII.7.B. Unsolved problem. Does the intersection of a decreasing sequence of 2-cells have the fixed-point property?

Some consider the above as one of the most interesting unsolved problems in plane topology. Discussions of it are found in §7 of [B_{22}] by Bing and Problem 107 in Mauldin [M_4].

That the problem is a planar question is shown by the following example in R^3.

VII.7.C. Fixed-point free example. Figure VII.7.C shows an example found by Bing [B_{22}] of a compact set X in R^3 which is an intersection of a decreasing sequence of 3-cells but for which there is a fixed-point free homeomorphisms of X onto itself.

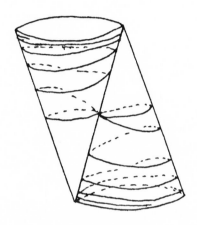

FIGURE VII.7.C.

The example is obtained by taking two cones (dunce caps) sewed together along a radial arc of each so that the vertex of each cone lies on the base of the other as shown in Figure VII.7.C. The top of the left cone is left invariant but the joining radial arc becomes a spiral and the vertex end of the right cone is wrapped around the left cone along this spiral. Similarly, the lower part of the left cone is wrapped around the lower part of the right cone. See §5 of [B_{22}].

For those who prefer explicit descriptions, it is suggested that the upper part of the left cone have an equation $1 \leqslant 1 + z = \sqrt{(x + 1)^2 + y^2} \leqslant 2$. Also, if $0 \leqslant z_0 < 1$, the plane $z = z_0$ intersects X in a small circle of radius $1 - z_0$ centered at the intersection of the plane $z = z_0$, the cylinder $(x + 1)^2 + y^2 = 4$, and such that the ray from $(-1, 0, 0)$ to the projection on the xy-plane of the center makes an angle of $1/(1 - z)$ with x-axis. The part of X below the xy plane consists of the reflection of the upper part through the origin.

The homeomorphism of X onto itself that moves each point raises each point between the planes $z = -1$ and $z = 1$ and raises them in such a way near these planes that continuity makes the end circles turn.   □

CHAPTER VIII

LINKING

In proving results about the intersection of surfaces with arcs or other surfaces it is convenient to understand the linking of simple closed curves. In a sense, the study of linking curves is a part of homology theory. It is not our purpose to cover homology theory but rather to concentrate on those topics of it of particular use in the study of 3-manifolds.

VIII. 1. Chains, cycles, and bounding cycles. We regard these as tools to help us understand the geometry of 3-manifolds rather than as objects of study.

A geometric complex is called an *i-chain* if it is the union of a finite number of *i*-simplexes. We designate an *i*-chain by C^i where *the superscript i designates the dimension of the chain*. Usually we shall not need a particular rectilinear triangulation T of C^i but if we do, we designate this with (C^i, T).

The mod 2 *boundary* of (C^i, T) is the union of the $(i - 1)$-simplexes of T that are faces of an odd number of *i*-simplexes of T. It follows from invariance of domain (Theorem VII.6.F) that if T_1, T_2 are two triangulations of the same C^i, then the mod 2 boundaries of (C^i, T_1), (C^i, T_2) are the same. Hence we use $\partial(C^i)$ to denote this mod 2 boundary and suppress the T. If $\partial C^i = C^{i-1}$, we say that C^{i-1} *bounds* C^i.

If $\partial(C^i) = \varnothing$, we call C^i an *i-cycle* and denote it by Z^i. We did not define the boundary of a 0-simplex so use the convention that a 0-*cycle* is a 0-*chain* with an even number of vertices.

If $(C_1^i, T_1), (C_2^i, T_2), (C_1^i \cup C_2^i, T_3)$ are triangulated complexes such that each of T_1, T_2 are subsets of T_3 the mod 2 *sum* of C_1^i and C_2^i (denoted by $C_1^i + C_2^i$) is the union of all *i*-simplexes of T_3 that belong to one but not both of T_1, T_2. Again, the mod 2 sum is independent of the triangulations.

It may be noted that chains, the mod 2-boundaries of chains, and the mod 2 sums of chains are point sets rather than collections of simplexes. Usually we suppose that chains lie in Euclidean spaces and inherit their metrics from these spaces.

THEOREM VIII.1.A. *If in R^3, Z^1, and C^1 are mutually disjoint, then there is a 2-chain C^2 such that $Z^1 = \partial(C^2)$ and C^2 is in general position with respect to C^1.*

PROOF. We suppose Z^1 and C^1 have given triangulations but we do not use notation for them. Let p be a point of R^3 which is not on any plane determined by three vertices of Z^1. Let C_0^2 be the cone over Z^1 from p and T_0 be a triangulation of C_0^2. By considering the complex (C_0^2, T_0) and pushing the vertices of T_0 not on Z^1 one at a time, one gets (C^2, T) where C^2 misses each vertex of C^1 and the 1-skeleton of T_0 misses C^1. □

The above theorem remains true if we replace R^3, Z^1, C^1 by R^m, Z^r, C^s $(r + s < m)$ and conclude that there is a C^{r+1}.

We do not claim that C^2 is unique. This is in contrast to how Z^2 bounds in R^3.

THEOREM VIII.1.B. *If Z^2 is a 2-cycle in R^3, there is one and only one 3-chain C^3 in r^3 such that $Z^2 = \partial(C^3)$.*

PROOF. To obtain C^3 one could start by finding a triangulation T_0 of r^3 such that $Z^2 \subset T_0^2$. One then finds a point p not in any plane determined by any three vertices of (Z^2, T_0) and cone Z^2 from p. Let $\{\sigma_i^3\}$ be the set of 3-simplexes each obtained as the join of p and a 2-simplex Δ_i^2 of T. It follows from a minor extension of Theorem I.1.A that there is a triangulation T_1 of R^3 such that each σ_i^3 is the union of simplexes of T_1.

Let C^3 be the union of the 3-simplexes of T_1 that lie in an odd number of σ_i^3's —that is, C^3 is the mod 2 sum of the σ_i^3's. Note that $\partial C^3 = C^2$. We shall show that Bd $C^3 = \partial C^3$.

Suppose Δ_1^3, Δ_2^3 are two elements of T_1 that share a common face Δ^2. If $\Delta^2 \subset Z^2$, it lies in a 2-simplex Δ_j^2 of T_0 and one (the one on the p side of Δ_j^2) but not both of Δ_1^3, Δ_2^3 lie in σ_i^3. For other σ_i^3, both or neither of Δ_1^3, Δ_2^3 lie in σ_i^3. Hence one but not both of Δ_1^3, Δ_2^3 lie in C^3 and $Z^2 \subset$ Bd C^3. This shows that $C^3 \neq 0$ if $Z^2 \neq 0$ and that indeed there is a C^3. If $\Delta^2 \not\subset Z^2$ and one but not both of Δ_1^3, Δ_2^3 lie in a σ_i^3, Δ^2 is on a plane R^2 through p. Since Z^2 is a cycle, the number of bases of these σ_i^3's is even, and there are the same number mod 2 of the σ_i^3's on the side of R^2 as on the other. Hence both or neither Δ_1^3, Δ_2^3 lie in C^3 and Bd $C^3 \subset Z^2$.

Now that we have shown the existence of a 3-chain C^3 so that $\delta(C^3) = Z^2$, we give an easier description of it. Let x be a point outside a 3-ball containing Z^2 and let C^3 be the closure of the set of all points p such that there is a PL arc px in general position with respect to Z^2 that crosses Z^2 an odd number of times.

In moving from p to x one would first cross out of C^3, then perhaps back in, then out again,..., and if one is to finally arrive at x, would have crossed Z^2 an odd number of times. We need to know of the existence of C^3 before using this simple approach to assure ourselves that there is not some PL arc from p to x that crosses Z^2 an odd number of times and another from p to x that crosses an even number.

It is this alternative description of C^3 that we use to show that C^3 is unique.

If (C_0^3, T) is a 3-chain in R^3, some 2-simplexes of T are faces of two 3-simplexes of T and some are faces of only one. The union of these that belong to only one is $\partial(C_0^3)$. This union is not only $\partial(C_0^3)$ but also Bd C_0^3. $\square$

Theorem VIII.1.B could be generalized by replacing R^3 and Z^2 by R^n and Z^{n-1} and concluding that there is one and only one C^n.

THEOREM VIII.1.C. *If a polygon P^1 bounds a 2-chain C^2 in R^3, then each neighborhood of C^2 contains a 2-mainfold-with-boundary M^2 such that* Bd $M^2 = P^1$.

PROOF. Let T be a triangulation of C^2. Already C^2 is locally a 2-manifold-with-boundary except possibly at the 1-skeleton of T. If Δ^1 is a 1-simplex of T that is an edge of more than two 2-simplexes of T we put a PL 3-ball B^3 about Δ^1 as shown in Figure VIII1.C. We require that $B^3 -$ Bd Δ^1 does not intersect any 2-simplex of T except those containing Δ^1. For each Δ_i^2 of T containing Δ^1, Bd $B^3 \cap \Delta_i^2$ is a disk that intersects Bd B^3 in an arc α_i^1. Suppose the α_i^1's are ordered $\alpha_1^1, \alpha_2^1, \ldots, \alpha_j^1$ about Bd B^3 in a natural fashion.

Remove $B^3 \cap C^2$ and put back the disk on Bd B^3 bounded by $\alpha_1 \cup \alpha_2$, the one bounded by $\alpha_3 \cup \alpha_4, \cdots$. If j is even this completes our first alteration, but if j is odd, we put back $\Delta_j^2 \cap B^3$. This modification is done for each Δ^1 on 3 or more faces of T of C^2, and we call the resulting 2-chain (C_1^2, T_1).

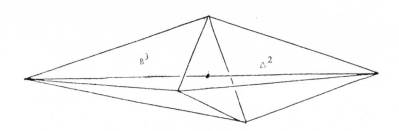

Figure VIII.1.C

Finally, we modify C_1^2 about the vertices of T_1. The star of such a vertex is the cone over some mutually disjoint simple closed curves and possibly an arc. If the closed curves are capped off (starting with an inner-most one) near the vertex, then a 2-chain meeting the requirements of Theorem VIII.1.C results. $\square$

Working in a 3-manifold-with-boundary rather than in R^3 gives the following modification of Theorem VIII.1.C.

THEOREM VIII.1.D. *If M^3 is a triangulated 3-manifold-with-boundary and P is a polygon on* Bd M^3 *that bounds in M_3, then there is PL 2-manifold-with-boundary M^2 in M^3 such that $P = $ Bd $M^2 = M^3 \cap $ Bd M^3. $\square$*

Applying Theorem VIII.1.D and noting the impossibility of having an arc with just one end, one gets the following result.

THEOREM VIII.1.E. *Suppose M^3 is a triangulated 3-manifold-with-boundary and P_1, P_2 are polygons on* Bd M^3 *whose intersection is a 1-point set at which P_1 and P_2 cross. Then at least one of P_1, P_2 does not bound in M^3.* □

VIII.2. Linking polygons. Suppose P_1^1, P_2^1 are two polygons in R^3. We say that P_1^1 *links* P_2^1 if and only if there is a 2-chain (C^2, T) in R^3 such that P_1^1 bounds C^2, (C^2, T) is in general position with respect to P_2^1 and $N(C^2, P_2^1)$ is *odd* where $N(A, B)$ denotes the number of points in $A \cap B$. In case $N(C^2, P_2^1)$ is odd (or even) we say that P_2^1 *crosses* C^2 an odd (or even) number of times.

We show that if P_1^1 links P_2^1 then every chain C^2 bounded by P_1^1 and in general position with respect to P_1^1 had odd intersection with P_2^1. Also, we show that P_1^1 links P_2^1 if and only if P_2^1 links P_1^1. To that end we let (C_1^2, T_1), (C_2^2, T_2) be 2-chains bounded by P_1^1 and (C_0^2, T_0) be a 2-chain bounded by P_2^1 and such that (C_0^2, T_0) is in general position with respect to each of (C_1^2, T_1), (C_2^2, T_2). Denote the 2-simplexes of T_1 by $(\sigma_1^2, \sigma_2^2, \ldots, \sigma_r^2)$ and those of T_0 by $(s_1^2, s_2^2, \ldots, s_k^2)$.

Using the fact that $N(\sigma_i^2, \text{Bd } s_j^2) = N(\text{Bd } \sigma_i^2, s_j^2) \bmod 2$ and some 1-simplexes are the faces of an even number of 2-simplexes, we find that the following relations hold mod 2:

$$N(C_1^2, P_2^1) = \sum N(\sigma_i^2, P_2^1)$$
$$= \sum \sum N(\sigma_i^2, \text{Bd } s_j^2) = \sum \sum N(\text{Bd } \sigma_i^2, s_j^2)$$
$$= \sum N(P_1^1, s_j^2) = N(P_1^1, C_0^2).$$

Similarly, $N(C_2^2, P_2^1) = N(P_1^1, C_0^2) \bmod 2$ so $N(C_1^2, P_2^1) = N(C_2^2, P_2^1) \bmod 2$ and whether or not P_1^1 links P_2^1 is independent of the C^2 chosen. Since $N(C_1^2, P_2^1) = N(C_0^2, P_1^1) \bmod 2$. P_1^1 links P_2^1 if and only if P_2^1 links P_1^1. Note that we permitted polygons to bound nonorientable surfaces.

THEOREM VIII.2.A. *If P_1^1, P_2^1 are two mutually disjoint polygons in R^3, P_1^1 links P_2^1 if and only if P_2^1 links P_1^1.* □

We have shown the following.

THEOREM VIII.2.B. *Suppose in R^3 that C_1^2, C_2^2 are two chains bounded by polygon P_1^1 and P_0^1 is a polygon which is GP with respect to C_1^2 and C_2^2. Then $N(C_1^1, P_0^1) = N(C_2^2, P_0^1) \bmod 2$.* □

THEOREM VIII.2.C. *If P_1^1, P_2^1 are disjoint polygons in R^3 that do not link, P_1^1 bounds a 2-chain in $R^3 - P_2^1$.*

PROOF. Let C^2 be a 2-chain in general position with respect to P_2^1 which bounds P_1^1. Since P_1 does not link P_2, $N(C^2, P_2^1)$ is even. We alter C^2 by finding two adjacent (with respect to order on P_2^1) points of $C^2 \cap P_2^1$, cutting small holes in C^2 about the holes, and running a tube centered on an arc of P_2^1 between the holes. If this operation is performed a finite number of times, one changes C^2 to a 2-chain in $R^3 - P_2^1$ that bounds P_1^1. □

We wait until §VIII.3 for a treatment of the case where the simple closed curves are not PL.

We use the following result in the next section to treat the linking of topological simple closed curves in R^3 rather than just PL ones.

THEOREM VIII.2.D. *Suppose* P_1^1, P_2^1, P_0^1 *are three polygons in* R^3 *such that* $P_1^1 + P_2^1$ *bounds in* $R^3 - P_0^1$. *Then* P_0^1 *links* P_1^1 *if and only if it links* P_2^1. $\square$

VIII.3. Linking curves. Suppose in R^3 that J_1, J_2 are two simple closed curves (not necessarily PL) and $d(J_1, J_2) > \varepsilon$. Let P_1^1, P_2^1 be PL curves in R^3 such that there is a homeomorphism h_i of J_i onto $P^1 i$ such that $d(h_i, \text{Id}) < \varepsilon/2$. We say that J_1 links J_2 if and only if P_1^1 links P_2^1. It follows from Theorem VIII.2.D that whether or not J_1 links J_2 is independent of the approximating P_1^1's chosen. Hence to investigate the linking of arbitrary simple closed curves, we investigate the linking of nearby polygons.

The following theorem is a consequence of Theorem VIII.2.C.

THEOREM VIII.3.A. *Suppose* J_1, J_2, J_0 *are simple closed curves in* R^3 *such that* J_1 *links* J_0 *and there is a homeomorphism* h_1 *of* J_1 *onto* J_2 *and a homotopy* $H(x, t)$: $J_1 \times [0, 1] \to R^3 - J_0$ *such that* $H(x, 0) = x$, $H(x, 1) = h_1(x)$. *Then* J_1 *links* J_0 *if and only if* J_2 *links* J_0. $\square$

In light of the following, we speak not of whether or not J_1 links J_2 but rather whether or not J_1, J_2 link each other.

THEOREM VIII.3.B. *In* R^3 *simple closed curve* J_1 *links a simple closed curve* J_2 *if and only if* J_2 *links* J_1. $\square$

THEOREM VIII.3.C. *Suppose* θ *is a* θ-*curve in* R^3 *and* J *is a simple closed curve in* $R^3 - \theta$. *Then* J *links exactly two simple closed curves in* θ *if it links one of them.*

PROOF. We use Theorem VII.3.A and suppose θ is the union of three PL arcs pxq, pyq, pzq, and J is PL. Let C^2 be a 2-chain which is in general position with respect to pxq, pyq, pzq, and which is bounded by J. Suppose J links $pxq \cup pyq$. Then one of $N(pxq, C^2)$, $N(pyq, C^2)$ is odd, (say pxq) and the other (say pyq) is even. If $N(pzq, C^2)$ is even, $pxq \cup pzq$ links J and $pyq \cup pzq$ does not. Also, the theorem follows with the roles of pxq, pyq reversed if $N(pzq, C^2)$ is odd. $\square$

VIII.4. The Alexander addition theorem. It is sometimes easier to prove things about PL objects than about topological objects. One of the tools for inferring topological truths from PL results is the Alexander addition theorem. This theorem has been called one of the most useful theorems in the study of Euclidean spaces. Wilder has an excellent treatment of the theorem in §5 of Chapter II of [W_7].

THEOREM VIII.4.A (ALEXANDER ADDITION THEOREM). *Suppose in* R^n *that* A, B *are closed sets and* Z^{r-2}, C_1^{r-1}, C_2^{r-1}, C^r *are chains such that* Z^{r-2}

bounds C_1^{r-1} in $R^n - A$, Z^{r-2} bounds C_2^{r-1} in $R^n - B$, $C_1^{r-1} + C_2^{r-1}$ bounds C^r in $R^n - (A \cap B)$. Then there is an $(r-1)$-chain C^{r-1} such that Z^{r-2} bounds C^{r-1} in $R^n - (A \cup B)$.

PROOF. This theorem is an extension of Theorem III.4.A and follows from a variation of methods used there. There we removed a portion of a rectangular disk and had a set X left that missed B. Here we remove a part of C^r near $C^r \cap B$ and have left an r chain C_1^r that contains $C_2^{r-1} \cup (C^r \cap A)$ but misses $C^r \cap B$. The required chain $C^{r-1} = C_2^{r-1} + \partial C_1^r$. See Figure VIII.4.A.

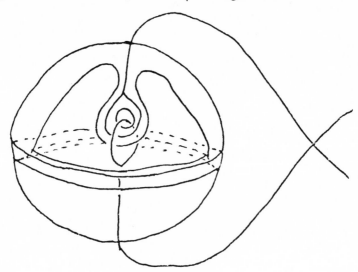

Figure VIII.4.A

Suppose C_1^{r-1} is adjusted so that $C_1^{r-1} \subset C_2^{r-1} \cup C_2^{r-1}$. To get a more precise description of C_1^r, we let T be a triangulation of R^n such that each of Z^{r-2}, C_1^{r-1}, C_2^{r-1}, C^r is the union of a finite number of simplexes of T and the mesh of T is less than $d(C_2^{r-1} \cup (C^r \cap A), C^r \cap B)/3$. Let C_1^r be the union of the simplexes of T that lie in C^r and intersect $C_2^{r-1} \cup (C^r \cap A)$. Then $C^{r-1} = \partial C_1^r + C_2^{r-1}$.

It may be noted that we picked C^{r-1} close to $C_2^{r-1} \cup (C^r \cap A)$. It may be noted that by picking the mesh of T even smaller we could have picked C^{r-1} even closer to $C_2^{r-1} \cup (C^r \cap A)$. Similarly, we could have picked it very close to $C_1^{r-1} \cup (C^n \cap B)$. $\square$

We present the following theorem as in interesting and easy application of the Alexander addition theorem.

THEOREM VIII.4.B. *Suppose A, B are mutually disjoint closed subsets of R^n and p, q are points of $R^n - (A \cup B)$ and neither A nor B separates p from q. Then $A \cup B$ does not either.*

PROOF. Let Z^0 be the 2-point set (p, q), C_1^1 be a PL arc in $R^n - A$ from p to q, C_2^1 be a similar arc in $R^n - B$, and C^2 be a 2-chain in R^n bounded by $C_1^1 + C_2^1$.

We suppose $n \geqslant 2$ and $C_1^1 \cup C_2^1$ is a nontrivial cycle. Then Theorem VIII.4.A promises a 1-chain C^1 in $R^n - (A \cup B)$ bounded by Z_0.   □

THEOREM VIII.4.C. *Theorem* VIII.4.A *remains true in the case* $n = r$ *if we replace the hypothesis that* $C^n \subset R^n - (A \cap B)$ *with the one that* $A \cap B \subset C^n$.

There are two ways of looking at the above variation of Theorem VIII.4.A. Either pick C_1^n to lie in $R^n - C^n$ and near $C_2^{n-1} \cup A$ or apply Theorem VIII.4.A after making $\overline{R^n - C^n}$ into an n-chain by putting in the point at infinity and removing another point.   □

VIII.5. Balls do not link. If A is a subset of S^n and Z^r is an r-cycle in $S^n - A$ we say that Z^r *links* A if it does not bound in $R^n - A$. This definition is not complete in that it speaks of a PL cycle. A critical eye might note that we have not yet shown that this second definition of linking (the first appearing in §VIII.3) agrees in case Z^r is a polygon P^1 in R^3 and A is a simple closed curve disjoint from P^1. Would it follow that P^1 bounds in the complement of A if it bounds in the complement of a close approximation of A. We wait for Theorem VIII.6.B before giving an affirmative answer.

It follows from the following application of the Alexander addition theorem that if P^1 is a polygon in R^3 and B is a topological ball in $R^3 - P^1$, then P^1 bounds in $R^3 - B$.

THEOREM VIII.5.A. *Suppose in* R^n $(n \geqslant 2)$ *that* C^{r-1} $(r \leqslant n)$ *is an* $(r - 1)$-*chain bounded by an* $(r - 2)$-*cycle* Z^{r-2} *and* B^i $(i = 0, 1, \ldots, n)$ *is a topological i-ball in* $R^n - Z^{r-2}$. *Then there is* $(r - 1)$-*chain* C_0^{r-1} *in* $R^n - B^i$ *that bounds* Z^{r-2}.

PROOF. Proof is by induction on i. In case $i = 0$, we get C_0^{r-1} by moving C^{r-1} off a one-point set B^0.

It we want to establish the theorem for $i = i_0$ and have already established it for lesser values, we express B^{i_0} as union of i_0-balls $B_1, B_2, \ldots, B_m$ so that each $B_{k+1} \cap (\cup_{j \leqslant k} B_j)$ is an $(i_0 - 1)$-ball, and each B_j is so small that Z^{r-2} bounds in $R^n - B_j$. Suppose Z^{r-2} bounds K_1^{r-1} in $R^n - (\cup_{j \leqslant k} B_j)$ and bounds K_2^{r-1} in $R^n - B_{j+1}$. We conclude from Theorems VIII.4.A or VIII.4.C that Z^{r-2} bounds K^{r-1} in $R^n - (\cup_{j \leqslant k+1} B_k)$. Induction shows that there is an $(r - 1)$-chain C_k^{r-2} bounded by Z^{r-2} and missing $B_1 \cup B_2 \cup \cdots \cup B_k$. Hence there is a $C_0^{r-1} = C_m^{r-1}$ in $R^n - B^{i_0}$ which is bounded by Z^{r-2}.   □

Adding "the point of infinity" converts R^n to S^n. Each $(n - 1)$-cycle in R^n which bounds an n-chain C^n also bounds $\overline{S^n - C^n}$. Hence, a change from R^n to S^n loses the uniqueness property of Theorem VIII.1.B but gained a dimension for Theorem VIII.5.A. Theorem VIII.5.A would have been false if we had replaced "$(r \leqslant n)$" by "$(r \leqslant n + 1)$".

THEOREM VIII.5.B. *If* K^2 *is a 2-sphere in* R^3 *and* P^1 *is a polygon in* $R^3 - K^2$, *then* P^1 *bounds in* $R^3 - K^2$.

PROOF. Express K^2 as the union of disks D_1, D_2 so that $D_1 \cap D_2$ is a simple closed curve J. Then P_1 bounds a 2-complex C_i^2 in $R^3 - D_i$ ($i = 1, 2$), and $C_1^2 + C_2^2$ bounds a 3-complex C^3. We suppose $C_i^2 \subset C_1^2 + C_2^2$. If $J \cap C^3 = \varnothing$ we get the 2-complex C^2 bounded by P^1 by digging into C^3 near one of D_1, D_2. If $J \subset C^3$, we get C^2 by digging into $R^3 - C^3$. $\square$

An open subset U of R^n is n-*connected in the* mod 2 *homology sense* if each n-cycle in U bounds in U. It is *locally n-connected* (n-lc) at a point $p \in U$ if for each neighborhood N_1 of p there is a neighborhood N_2 of p such that each n-cycle in $U \cap N_2$ bounds in $U \cap N_1$. It is *uniformly locally n-connected* (n-ulc) if for each $\varepsilon > 0$ there is a $\delta > 0$ such that each n-cycle of diameter less than δ in U bounds an $(n + 1)$-chain in U of diameter less than ε. It may be noted that in homology one uses such abbreviations as n-ulc (or n-lc) instead of the abbreviations n-ULC or (n-LC) used in homotopy.

If one adjust the proof of Theorem VIII.5.B by taking D_1 and C_1^2 small and taking C^2 near $D_1 + C_1^2$, one obtains the following.

THEOREM VIII.5.C. *If K^2 is a 2-sphere in R^3, each component of $R^3 - K^2$ is* 1-*ulc.* $\square$

THEOREM VIII.5.D. *Suppose in S^n ($n \geqslant 2$) that C^r ($r \leqslant n$) is an r-chain bounded by an $(r - 1)$-cycle Z^{r-1} and B^i ($i = 0, 1, \ldots, n$) is a topological i-ball in $S^n - Z^{r-1}$. Then there is an r-chain C_0^r in $S^n - B^i$ that bounds Z^{r-1}.* $\square$

The statements of the last two theorems speak of the existence of chains but do not tell where they are. The following theorem give a possible location.

THEOREM VIII.5.E. *The $(r - 1)$-chain C_0^{r-1} promised by Theorem VIII.5.A can be chosen to lie in any open set containing $\overline{U} \cup B^i$ where U is the union of components of $C^{r-1} - B^i$ intersecting Z^{r-2}.*

PROOF. Let V be an open set containing $\overline{U} \cup B^i$. Since there is a retraction of $C^{r-1} - U$ into B^i, we suppose with no loss of generality that $C^{r-1} \subset V$.

The K_1^{r-1} and K_2^{r-1} are picked to lie in V but we exercise little control over the r-chain K^r bounded by $K_1^{r-1} + K_2^{r-1}$ except that it misses or contains $B_{k+1} \cap (\cup_{j < k} B_j)$. However, the K^{r-1} in C^r (or $\overline{R^n - C^r}$) promised by Theorem VIII.5.A or VIII.5.C is close to $K_2^{r-1} \cup (\cup_{j < k} B_j)$. This makes the resulting C_0^{r-1} lie close to $C_2^{r-1} \cup B^{i_0}$. $\square$

In particular cases a person might make an even better approximation to a possible location at C_0^{r-1}. It may be shown that in R^3 if B^2 is a topological 2-ball and pq is an arc from whose ends miss B^2 then there is an arc rs in B^2 such that for any neighborhood of $pq \cup rs$ there is an arc from p to q in $N - B^2$. We do not pursue this but mention that the methods of proof reveal more than the stated results.

VIII.6. Nonlinking curves. It is the purpose of this section to show that if two mutually disjoint simple closed curves J_1, J_2 in R^3 fail to link, then J_1 can be approximated by a polygon P_1 that bounds in the complement of J_2. This theorem is an extension of Theorem VIII.2.B but one should not fall for the trap of supposing that if a 2-chain C^2 misses an approximation P^1 to J, then C^2 misses J.

THEOREM VIII.6.A. *Suppose in R^3 that J_1, J_2 are two mutually disjoint simple closed curves J_1, J_2 which do not link. Then J_1 can be approximated by a polygon P^1 such that P^1 bounds a 2-chain C^2 in $R^3 - J_2$.*

PROOF. A clue to proving the theorem is to think of J_2 not as a simple closed curve but as the union of two arcs pxq, pyq and recall that 1-cycles do not link arcs in R^3 (Theorem VIII.5.A).

Let P^1 be a close PL approximation to J_1 and C_1^1, C_2^2 be 2 chains in $R^3 - pxy$, $R^3 - pyz$ bounded by P^1. Theorem VIII.5.A guarantees such chains. Let C^3 be the 3-chain in R^3 bounded by $C_1^2 + C_2^2$. Let $px'q$, $py'q$ be close PL approximations to pxq, pyq such that $px'q \cup py'q$ is a polygon in general position with respect to $\partial(C^3)$. Since J_1 and J_2 do not link, $N(C_1^2, px'q)$ is even and either both or neither of p, q lie in C^3.

If neither p nor q lie in C^3, we let A, $B = pxq$, pyq and quote Theorem VIII.4.A to obtain a C^2 in $R^3 - (A \cup B) = R^3 - J_2$ bounded by P^1. Recall that C^2 was obtained in proof of Theorem VIII.4.A to lie near $C_2^2 \cup pxq$. If both p and q lie in C^3, $C^2 \subset R^3$; if neither lies in C^3, $C^2 \subset R^3 - C^3$. In a sense, we have shifted the point at infinity and used $R^3 - C^3$ instead of C^3. $\square$

Our proof of Theorem VIII.6.A also shows the following.

THEOREM VIII.6.B. *Suppose that J is a simple closed curve in R^3 and P^1 is a polygon in $R^3 - J$ that does not link J. Then P^1 bounds a 2-chain C^2 in $R^3 - J$.* $\square$

VIII.7. Homology groups. In Chapter V we used equivalence classes of loops to describe the fundamental group π_1. In a similar fashion equivalence classes of cycles can be used to describe homology groups.

Suppose U is a space (such as an open subset of a polyhedron) where there are chains, cycles, and bounding cycles. We denote the mod 2 nth homology group of U by $H_n(U, Z_2)$ and define it as follows: An element of $H_n(U, Z_2)$ is an equivalence class of n-cycles in U such that two n-cycles Z_1, Z_2 belong to the same equivalence class if and only if $Z_1 + Z_2$ bounds in U; the product of two elements γ_1, γ_2 of $H_n(U, Z_2)$ is the equivalence class of $H_n(U, Z_2)$ containing $Z_1 + Z_2$ where $Z_1 \in \gamma_1$, $Z_2 \in \gamma_2$. Note that the identity element of $H_n(U, Z_2)$ is the set of bounding cycles in U.

Wilder has an excellent treatment of mod 2 homology in [$\mathbf{W_7}$]. Homology groups and variations of them have been useful tools for discovering geometric properties by algebraic techniques. It seems that they are used more extensively by those with algebraic tendencies whose interests often turn to a study of the tools rather than just their geometric implications. We refer to $H_n(U, Z_2)$ in §§IX.3, XV.4, XV.5.

CHAPTER IX

SEPARATION

Intuitively we say that a topological 2-sphere K^2 in R^3 separates p from q in R^3 if an arc from p to q crosses K^2 an odd number of times. This is a natural concept and has an advantage over the usual way of dealing with generators of groups. It has the disadvantage of the difficulty of seeing whether an arc crosses a topological surface an odd number of times. Rather than abandoning the number-of-crossings concept, we shall give it a precise meaning.

Although the results of this chapter extend to other dimensions we only treat them in the case $n = 3$. Unless we specify to the contrary, it is assumed that sets being discussed lie in R^3.

IX.1. General position approximations. We used the notion of general position in §IV.7 to consider whether the low-dimensional simplexes of one complex (X_1, T_1) intersected those of another (X_2, T_2). The *complexes were GP in R^3* if $T_1^i \cap T_2^j = \varnothing$ when $i + j < 3$. One could not simplify $X_1 \cap X_2$ with a slight push on the vertices of T_1.

One also uses the notion of general position to restrict the complexities of singularities. A *linear map f of a complex (X, T) into R^3 is GP* if for simplexes Δ^i, Δ^j of T with dimensions i, j,

$$\text{dimension} \left[f(\Delta^i) \cap f(\Delta^3) - f(\Delta^i \cap \Delta^j) \right] \leqslant i + j - 3,$$

and f is a homeomorphism on Δ^i.

If the dimension of X is not more than 3, we can drop the restriction that f is a homeomorphism on Δ^i.

A map of a geometric complex X into R^3 is a GP map if there is a triangulation T of X such that f on (X, T) is a GP map. If X is a geometric complex and h is a homeomorphism of X into $h(X)$, a map f of $h(X)$ is a GP map if fh is GP. If $d(f, \text{Id}) \leqslant \varepsilon$, we call f a GP ε-approximation of $h(X)$. If ε is small, we call f a GP close approximation of $h(X)$.

Before making further use of general position we comment on some deficiencies with what we are doing. In Figure VI.3.a we wished to avoid the unnecessary intersections of triples. In some studies (see Bing [B_{28}]) we may wish to avoid unnecessary intersections of triples, quadruples,..., n-tuples,.... However, in this book we usually need to concern ourselves only with the intersection of pairs and tailored our definitions accordingly. Had we been frequently concerned with triples we would have restricted the dimension of $f(\Delta^i) \cap f(\Delta^j) \cap f(\Delta^j)$. We do not consider the definition of GP as set in concrete.

Suppose S is a 2-sphere in R^3 and p, q are two points of $R^3 - S$. We use general position to determine whether or not S separates p from q in S. Let pq be a PL arc in R^3 from p to q. Then S separates p from q if and only if pq crosses S an odd number of times—but what does this mean?

To determine whether or not pq crosses S an odd number of times we let f be a GP ε-approximation of S where ε is less than the distance between S and $\{p\} \cup \{q\}$. Then $f(S)$ is a singular 2-sphere the image of whose set of singularities is of dimension 1. Adjust pq so that pq is GP with respect to $f(S)$. We say that pq crosses S an odd number of times if and only if the number of points in $pq \cap f(S)$ is odd. We shall show that whether or not pq crosses S an odd number of times is independent of the GP ε-approximation f of S and the PL arc pq from p to q. Also, there is an arc from p to q in $R^3 - S$ if and only if pq crosses S an even number of times.

As a start, we consider the following results.

THEOREM IX.1.A. *If Δ^3 is a 3-simplex and pq is a PL arc such that pq and Δ^3 are GP in R^3, then pq crosses* Bd Δ^3 *an odd number of times if and only if p, q lie in different components of $R^3 -$ Bd Δ^3.*   □

THEOREM IX.1.B. *Suppose in R^3 that Z^0 is a 0-cycle, S is a 2-sphere, and $\varepsilon < d(Z^0, S)$. If f_i ($i = 1$ or 2) is a GP ε-approximation of S and Z^1 is a 1-chain bounded by Z^0 such that Z^1 and $f_i(S)$ are GP, then Z^1 crosses $f_1(S)$ an odd number of times if and only if it crosses $f_2(S)$ an odd number of times.*

PROOF. In §VI.7 we changed one PL loop to another in a finite number of moves each of which consisted of replacing a part of the boundary of a 2-simplex with the complementary part. In a similar fashion we can change $f_1(S)$ to $f_2(S)$ by moves $f_1(S) = C_1^2, C_2^2, \ldots, C_m^2 = f_2(S)$ where each $C_i^2 + C_{i+1}^2$ is the boundary of a 3-simplex, the 3-simplex misses Z^0, and each C_i^2 and Z^1 are GP. It follows from Theorem IX.1.A that $Z^1 \cap C_i^2$ had an odd number of points if and only if $Z^1 \cap C_{i+1}^2$ does. Theorem IX.1.B follows.   □

That the number mod 2 of times that pq crosses $f_i(S)$ is independent of the PL arc pq chosen is shown by the following result.

THEOREM IX.1.C. *Suppose Z^2 is a 2-cycle and P^1 is a polygon such that Z^2 and P_1^1 are GP in R^3. Then the number of points in $P^1 \cap Z^2$ is even.*   □

IX.2. Separation by spheres. Suppose K^2 is a topological 2-sphere in R^3 and p, q are points of $R^3 - K^2$. We show that K^2 separates p from q if and only if some arc from p to q crosses K^2 an odd number of times.

THEOREM IX.2.A. *The topological 2-sphere K^2 separates p from q in R^3 if some arc pq from p to q crosses K^2 an odd number of times.*

PROOF. Suppose with no loss of generality that pq is PL. If K^2 did not separate p from q, pq is an arc in a polygon P^1 such that $P^1 \cap K^2 \subset pq$. We show that the assumption that there is such a P^1 leads to a contradiction.

Let D^2 be a 2-ball in K^2 so that $K^2 \cap pq \subset \text{Int } D^2$. Let $h \colon K^2 \to S^2$ be a homeomorphism of K^2 onto a PL 2-sphere and $f \colon S^2 \to R^3$ be a GP approximation that is close to h^{-1}. Then $fh^{-1} \colon K^2 \to R^3$ is a GP map that is close to the identity. Suppose $gh^{-1}(\text{Bd } D^2)$ is a simple closed curve. Pick P^1 in GP with respect to each of $gh^{-1}(D^2)$ and $gh^{-1}(K^2 - \text{Int } D^2)$. But

$$N(P^1, gh^{-1}(K^2 - \text{Int } D^2)) = \varnothing$$

and $N(P^1, gh^{-1}(D^2))$ is odd. This contradicts Theorem VIII.2.A. $\square$

THEOREM IX.2.B. *A topological 2-sphere K^2 fails to separate p from q if some arc pq crosses K^2 an even number of times.*

PROOF. Suppose without loss of generality that pq is PL. Let D^2 be a 2-ball in K^2 so that $K^2 \cap pq \subset \text{Int } D^2$. Let pxq be a PL arc from p to q in $R^3 - D^2$ so that $pq \cup pxq$ is a polygon P^1. This polygon P^1 fails to link Bd D^2 since it crosses D^2 an even number of times. Hence it follows from Theorem VIII.6.B that P^1 bounds a 2-chain C^2 in $R^3 - \text{Bd } D^2$. However, the Alexander addition theorem says that since pq misses $K^2 - \text{Int } D^2$, pxq misses D^2, and C^2 misses $D^2 \cap (K^2 - \text{Int } D)$, then there is a pyq that misses K^2. But pyq shows that K^2 does not separate p from q. $\square$

Where is the arc pyq? The following theorem provides some relevant information.

THEOREM IX.2.C. *Suppose K^2 is a 2-sphere, pq is an arc that crosses K^2 an even number of times, and D^2 is a 2-ball in K^2 that contains $pq \cap K^2$. Then each neighborhood of $pq \cup D^2$ contains an arc pyq in $R^3 - K^3$ from p to q.*

PROOF. We only need to examine the construction of pyq from the proof of Theorem VIII.4.A. Once we have obtained C^2, we keep a part C_1^2 of it near $pq \cup D^2$ as done in proof of Theorem VIII.4.A. Since C_1^2 lies near $pq \cup D^2$, pyq does also. $\square$

IX.3. Jordan-Brouwer theorem. The proof of the Jordan-Brouwer theorem we gave for the plane in Chapter III was given—not because it generalized but because it was easy. The procedure that we use here does generalize.

We showed in §IX.2 that if K^2 is a 2-sphere in R^3 and p, q are points of

$R^3 - K^2$, then K^2 separates p from q if and only if some arc from p to q crosses K^2 an odd number of times.

THEOREM IX.3.A. (JORDAN-BROUWER THEOREM FOR R^3). *If K^2 is a topological 2-sphere in R^3, then $R^3 - K^2$ has exactly two components and K^2 is the boundary of each.*

PROOF. There are several ways of showing that K^2 separates R^3. One uses the fact that if K^2 does not, then there is a retraction r of R^3 onto K^2. See Theorem VII.3.B. Let h be a homeomorphism of Bd Δ^2 onto K^2 and f be an extension taking Δ^2 into R^3. Then $h^{-1}rf$ is a retraction of Δ^2 onto Bd Δ^2—but this contradicts Theorem VII.1.A.

Since no 2-ball separates R^3 (Theorem VII.6.D), each component of $R^3 - K^2$ has K^2 as a boundary.

If p and q_i ($i = 1, 2$) belong to different components of $R^3 - K^2$, then there is an arc from q_1 to q_2 that crosses K^2 an even number of times. It follows from Theorem IX.2.A that q_1 and q_2 belong to some component of $R^3 - K^2$. Hence, $R^3 - K^2$ has only two components. □

One can get Theorem IX.3.A by proving a much more general theorem and then getting Theorem IX.3.A as a corollary. For example, if one uses $r_p(L)$ to denote the minimum number of generators in the pth homology mod 2 of L and Σ^s to denote a topological s-sphere in S^n, the following result is sometimes proved. See, for example, Hocking and Young [**HY**, p. 361] or Wilder [**W$_6$**, p. 61].

THEOREM IX.3.B. *If $p \neq k$,*

$$r_p(\Sigma^k) = r_{n-p-1}(S^n - \Sigma^k) = 0.$$

$$r_k(\Sigma^k) = r_{n-k-1}(S^n - \Sigma^k) = 1. \quad \Box$$

We do not follow this route. While it leads to results on spheres, our goal is to work at unlocking the mysteries of R^3. The following result, however, is worthy of note.

THEOREM IX.3.C (JORDAN-BROUWER THEOREM FOR R^n). *If K^{n-1} is a topological $(n-1)$-sphere in R^n, then $R^n - K^{n-1}$ has exactly two components and K^{n-1} is the boundary of each.* □

A set is called an *ε-set* if its diameter is less than ε.

A set X is *r-ULC* (*uniformly locally r-connected*) if for each $\varepsilon > 0$ there is a $\delta > 0$ such that any map of Bd I^{r+1} into a δ-subset of X can be extended to map I^{r+1} into an ε-subset of X.

THEOREM IX.3.D. *If K^2 is a 2-sphere in R^3, each component of $R^3 - K^2$ is 0-ULC.*

PROOF. Let δ be a positive number so small that any δ-subset of K^2 lies in an $\varepsilon/3$-disk in K^2. Let p, q be two points belonging to the same components of $R^3 - K^2$ and pq be a δ-arc from p to q. Suppose pq crosses K^2 an even number of times and D^2 is an $\varepsilon/3$-disk in K^2 containing $pq \cap K$. It follows from Theorem IX.2.C that in any neighborhood V of $pq \cup D^2$ there is an arc pyq in $R^3 - K^2$ from p to q. For the neighborhood sufficiently close to $pq \cup D^2$, pyq is an ε-arc.   □

IX.4. Local separation. An arc pq *crosses a disk* D^2 *an odd (or even) number of times* according to $N(W^2, p'q')$ is odd (or even) where $(W^2, p'q')$ is a close GP approximation of (D^2, pq) such that W^2 and $p'q'$ are GP with respect to each other. The following theorem shows that a disk locally separates R^3.

THEOREM IX.4.A. *If* D^2 *is a disk in* R^3 *and* $p_0 \in \text{Int } D^2$, *then there is an* $\varepsilon > 0$ *such that for any neighborhood* U *of* p_0 *there are two points* $p, q \in U - D^2$ *such that any* ε-arc *from* p *to* q *crosses* D^2 *an odd number of times.*

PROOF. Let $\varepsilon = \frac{1}{3} d(p_0, \text{Bd } D^2)$. Express Bd D^2 as the union of two arcs axb, ayb, and let ap_0b be a spanning arc of D^2. We show that there is a small simple closed curve p^1 in U that links $ap_0b \cup axb$ and the two points, p, q on p^1 that cannot be joined by a small arc in $R^3 - D^2$.

Let Δ^3 be a 3-simplex of diameter less than ε in U such that $p_0 \in \text{Int } \Delta^3$ and E^2 be a PL disk on Bd Δ^3 such that $ap_0 \cap \text{Bd } \Delta^3 \subset \text{Int } E$ and $(p_0b \cap \text{Bd } \Delta^3) \cap E^2 = \varnothing$. Then Bd $E^2 = P^1$. Since $a \in \text{Ext } \Delta^3$, $p_0 \in \text{Int } \Delta^3$, ap_0 crosses E^2 an odd number of times, and Bd E^2 links $ap_0b \cup axb$. Also, $ap_0b \cup axb$ links P^1 so P^1 crosses the subdisk D_0^2 of D^2 bounded by $ap_0b \cup a \times b$ an odd number of times.

Express Bd E^2 as the union of arcs $r_1r_2r_2r_3, \ldots, r_mr_1$ so that the r's do not lie on D^2 and no r_ir_{i+1} intersects both components of $D^2 - ap_0b$. Since Bd E^2 links Bd D_0^2, one of the r_ir_{i+1}'s crosses D_0^2 of odd number of times.

Assume there is an ε-arc in r_izr_{i+1} crossing D^2 an even number of times. We suppose that it together with r_ir_{i+1} makes a simple closed curve linking Bd D^2. However, $r_izr_{i+1} \cup r_ir_{i+1}$ has such small diameter that it can shrink to a point in $R^3 - \text{Bd } D^2$. Hence r_i, r_{i+1} can serve as the p, q guaranteed by the theorem.   □

The preceding theorem with the following shows that a disk D^2 locally has exactly two sides at each point of Int D^2.

THEOREM IX.4.B. *If* D^2 *is a topological disk,* $p_0 \in \text{Int } D^2$, *and* $\varepsilon > 0$, *there is a neighborhood* U *of* p_0 *such that if* p_1, p_2, p_3 *are three points of* $U - D^2$, *then some* ε-arc *in* $R^3 - D^2$ *joins some two of them.*

PROOF. Let D_0^2 be an $\varepsilon/3$-disk in D^2 with $p \in \text{Int } D_0^2$. Let U be a round open ball about p_0 which misses $D^2 - \text{Int } D_0^2$. Let p_1p_2, p_1p_3, p_2p_3 be straight arcs in U. Each of them misses $D^2 - \text{Int } D_0^2$ and at least one of them (say p_1p_2) crosses D_0^2 an even number of times. Let p_1xp_2 be an arc from p_1 to p_2 in $R^3 - D_0^2$ so that $p_1p_2 \cup p_1xp_2$ is a polygon P^1. Then P^1 does not link Bd D^2 and P^1 bounds a 2-chain C^2 in $R^3 - \text{Bd } D_0^2$ by Theorem VIII.6.A. Since p_1xp_2 misses D_0^2, p_1p_2 misses $D^2 - \text{Int } D_0^2$ and $p_1xp_2 \cup p_1p_2$ bounds C^2 in $R^3 - \text{Bd } D_0$, it follows from the Alexander addition theorem that there is an arc p_1zp_2 in $C^2 - D_2$. It follows from the proof of Theorem VIII.2.B that p_1zp_2 can be chosen to lie close to $p_1p_2 \cup D_0^2$ so it has diameter less than ε.   □

CHAPTER X

PULLING BACK FEELERS

In this chapter we are primarily concerned with R^3, and it is to be understood that this is the space in which we are working. (We do mention S^3 in §X.4 and in the proof of Theorem X.1.A.)

We discuss a procedure which even though relatively simple, has had a profound impact of what we know about Euclidean 3-space. See papers by Bing [B_{27}], Daverman [D_1], and Lininger [L_1]. The procedure is illustrated by the following special case. Suppose K^3 is a crumpled cube whose boundary is a wild sphere $K^2 = \text{Bd } K^3$ as shown in Figure X.A. There D^2 is a flat disk whose boundary misses K^3 but Bd D^2 cannot be shrunk to a point in $R^3 - K^3$. We seek a reembedding of K^3 so that Bd D^2 can be shrunk to a point in the complement of the modified K^3. The clue is given in Figure X.B where a blister is raised on each side of D^2. We show D^2 as horizontal so as to show the blisters. The part of K^3 that reaches down through D^2 is pushed up into the upper blister and the part of K^3 that reaches up through D^2 is pushed into the lower blister. It is to be admitted that during the homotopy pulling part of K^3 down and pushing part of it up that we do not have a homeomorphism at each intermediate stage but we do have one at the end and that is our goal.

X.1. Pull-back theorems. We generalize the procedure we alluded to in the preceding paragraph. While the result has its most important known applications in the case where X is a 2-sphere or crumpled cube, we state it in a more general setting.

THEOREM X.1.A. *Suppose that*
D^2 *is a round planar disk,*
V *is an open set containing* Int D^2,
X *is a compact locally connected subset of* $R^3 - \text{Bd } D^2$ *such that each loop in* X *can be shrunk to a point in* $R^3 - \text{Bd } D^2$, *and*
M *is a continuum in* $X - D^2$.

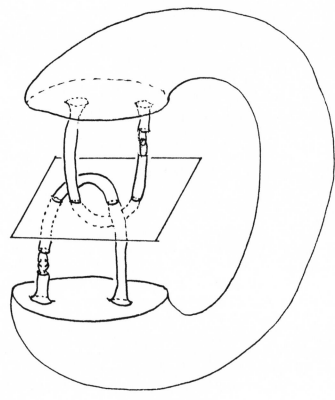

Figure X.A

Then there is a homeomorphism h of X into $(X \cup V) - D^2$ such that
   h is fixed on M,
   any point that is moved at all is moved into $V - D^2$, and
   if $p \in X - V$ lies in the component of X with M and h moves p into V, then Int D^2 *separates p from M in X.*

If X were a tame 2-sphere which intersected D^2 in a finite number of simple closed curves, we would be tempted to prove the theorem by cutting holes in X near D^2 and capping off these holes with disks in V to the side of D. This is a common technique that has been used with success. However, we shall use the theorem in a more general setting.

We regard X as the union of M and some feelers reaching out from M and poking through Int D^2. We even permit loops in the feelers to encircle other feelers but do not allow loops in the feelers to link Bd D^2.

The homeomorphism h given by Theorem X.1.A is said to result from *pulling back feelers* of X through Int D into V.

PROOF OF THEOREM X.1.A. One adds the point at infinity to R^3 to change it to a 3-sphere S^3. Note that $S^3 - $ Bd D^2 is an open solid torus and there is a

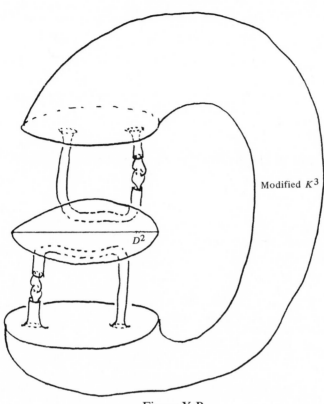

Modified K^3

D^2

Figure X.B

homeomorphism of $S^3 -$ Bd D^2 onto $S^1 \times$ Int D^2. We use $(\cos\theta, \sin\theta)$ to de-note a point of S^1 and designate $(\cos 0, \sin 0)$ by $p_0 \in S^1$. We pick the homeomor-phism $f\colon (S^3 -$ Bd $D^2) \to S^1 \times$ Int D^2 so that for $y \in$ Int $D^2, f(y) = (p_0, y) \in S^1 \times$ Int D^2. Let $R^1 \times$ Int D^2 be the universal covering space of $S^1 \times$ Int D^2 with the natural projection ρ: i.e., $\rho(x, y) = ((\cos x, \sin x), y)$.

Since each loop in X can be shrunk to a point in $R^3 -$ Bd D^2, $f(X)$ can be lifted—there is a map g of $f(X)$ into $R^1 \times$ Int D such that ρg is the identity on $f(X)$. We suppose that $gf(M) \subset (0, 2\pi) \times$ Int D^2. See Figure X.1.A.a.

It is convenient to regard V as a slight thickening of Int D^2 that misses M and $\rho^{-1}f(V) \cap [0, 2\pi] \times$ Int D^2 as the union of thin blisters on the ends of $[0, 2\pi] \times$ Int D^2. Let r be a homeomorphism of $R^1 \times$ Int D^2 onto $(0, 2\pi) \times$ Int D^2 that shoves $(-\infty, 0] \times$ Int D^2 into the left blister and $[2\pi, \infty) \times$ Int D^2 into the right blister. We require that r be the identity on $gf(M)$ and that it send any point that moves at all into $\rho^{-1}f(V)$. The required homeomorphism h is $f^{-1}\rho rgf$. The lower part of Figure X.1.A.a suggests how loops can be unhooked. $\square$

The homeomorphism h we have described satisfies a condition a bit more restrictive than we gave in the statement of Theorem X.1.A. If $p \in X - V$, $f(p) \in V$, and A is an arc from p to M in X, then not only is it true that A intersects D^2 but it must indeed cross D^2 algebraically a nonzero number of times. An open disk locally has two sides. In determining the number of times

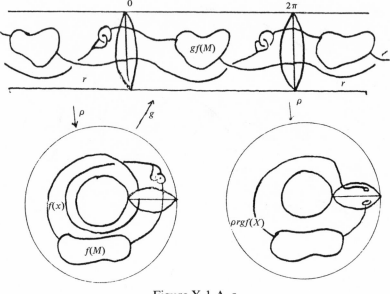

Figure X.1.A.a

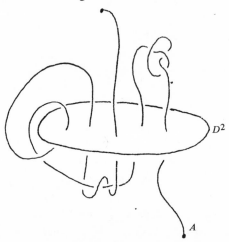

Figure X.1.A.b

that an arc A crosses a disk D^2, one can cancel out a crossing from one side to a second if it is paired with a crossing from the second. The arc A shown in Figure X.1.A.b crosses D^2 once algebraically. This notion of algebraic crossing is more elementary than the customary one using orientability, the right-hand rule, etc.

A disk D^2 in a 3-ball B^3 *spans* B^3 if $D^2 \cap$ Bd $B^3 =$ Bd D^2. We call D^2 *tame* if there is a homeomorphism of B^3 onto itself taking D^2 onto a 2-simplex. Similarly B^3 is *tame* if there is a space homeomorphism taking R^3 onto a 3-simplex. In an application of Theorem X.1.A to be used Chapter XI we will want D^2 to be a spanning disk of a tame 3-ball B^3 so we state it in the following form.

THEOREM X.1.B. *Suppose that*
B^3 *is a PL 3-ball,*

D^2 *is a PL spanning disk of* B^3,

X is a compact locally connected continuum in $R^3 -$ Bd D^2 *such that each loop in X can be shrunk to a point in* $R^3 -$ Bd D^2, *and*

M is a continuum in $X -$ Int B^3.

Then there is a homeomorphism h of X into $(X \cup$ Int $B^3) - D^2$ *such that*

h is fixed on the component of $X -$ Int B^3 *containing M,*

any point that is moved is moved into Int $B^3 - D^3$, *and*

a point p of $X -$ Int B^3 *is moved if and only if some arc from p to M crosses* D^2 *algebraically a nonzero number of times.* $\square$

We now return to a form of Theorem X.1.A that we will use in the next section. Instead of pulling X off of a disk D^2, we shall want to pull it off a finite number of such disks one at a time. The ordering of the D_i^2's makes a difference in the homeomorphism h finally obtained as can be seen from Figure X.1.C. If in that figure one pulls the feelers off $D_1^2 \cup D_2^2$ by considering D_1^2 first and then D_2^2, the point p is moved. However, if one considers D_2^2 first and then D_1^2, p is not moved.

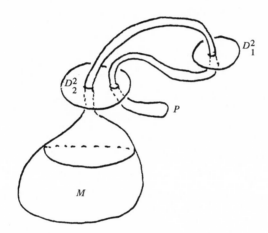

THEOREM X.1.C. *Suppose in* R^3 *that*

$D_1^2, D_2^2, \ldots, D_n^2$ *is a finite collection of disjoint tame disks.*

$V_1, V_2, \ldots, V_n$ *is a finite collection of disjoint open sets satisfying* Int $D_i^2 \subset V_i$,

X is a compact locally connected continuum in $R^3 - \cup$ Bd D_i^2 *such that each loop in X can be shrunk to a point in each* $R^3 -$ Bd D_i^2, *and*

M is a continuum $X - \cup D_i^2$. *Then there is a homeomorphism h of X into* $(X \cup \cup V_i) - \cup D_i^2$ *such that*

h is fixed on M,

any point that is moved at all is moved into some V_i,

if h moves $p \in X - V_i$, *then* D_i *separates p from M in X.* $\square$

We did not include in the statement of Theorem X.1.C how far h might move a point. Suppose each V_i had a diameter less than some ε_1 and each component of $X - D_i$ other than the one containing M had diameter less than ε_2. Then if $p \in V_i$

and $h(p) \in V_i$, $d(p, h(p)) < \varepsilon_1$. If $p \notin V_i$ but $h(p) \in V_i$, $d(p, D_i) < \varepsilon_2$ and $d(p, h(p)) < \varepsilon_2 + \varepsilon_1$.

A set is called an ε-*set* if its diameter is less than ε. A map f on a set X is called an ε-map if for each $x \in X$, $d(x, f(x)) < \varepsilon$. A map f is an ε-*approximation* to a map g if for each $x \in X$, $d(f(x), g(x)) < \varepsilon$.

X.2. Reimbedding a crumpled cube. As an application of Theorem X.1.C, we show how a crumpled cube can be pulled into the interior of a PL 2-sphere.

THEOREM X.2.A. *Suppose*

K^3 *is a crumpled cube*,

S^2 *is a PL 2-sphere*,

$\varepsilon > 0$,

M *is a continuum in* $K^3 \cap \text{Int } S^2$ *such that each component of* $K^3 - M$ *is an* ε-*set, and*

$D_1^2, D_2^2, \ldots, D_n^2$ *is a collection of mutually disjoint* ε-*disks on* S^2 *such that* $S^2 \cap K^2 \subset \bigcup \text{Int } D_i^2$.

Then there is a (2ε)-*homeomorphism* $h: K^3 \to \text{Int } S^2$ *such that* $h = identity$ *on* M. □

The side approximation theorem will not be treated until Chapter XIII, but we mention one version of it here since its statement is relevant to the statement of Theorem X.2.A and shows there is really an S^2 satisfying the hypotheses of that Theorem X.2.A.

THEOREM X.2.B. *Suppose* K^2 *is a 2-sphere (perhaps wild) in* R^3 *and* $\delta > 0$. *Then there is a* δ-*homeomorphism* h *of* K^2 *onto a PL 2-sphere* S^2 *in* R^3 *such that each component of* $K^2 - \text{Int } S^2$ *lies in a* δ-*disk in* K^2 *and each component of* $K^2 \cap S^2$ *lies in a* δ-*disk in* S^2. □

We say that S^2 *approximates* K^2 *almost* from Ext K^2. A similar statement says that there is a PL 2-sphere that approximates K^2 almost from Int K^2. If $S^2 \subset \text{Ext } K^2$, we would have dropped the "almost" and claimed that S^2 δ-*approximates* K^2 from the Ext K^2.

THEOREM X.2.C. *Suppose* K^3 *is a crumpled cube in* R^3 *and* $\varepsilon > 0$. *Then there is an* ε-*homeomorphism* $f: K^3 \to R^3$ *and an* ε-*homeomorphism* $h: \text{Bd } K^3 \to R^3$ *such that* $h(\text{Bd } K^3)$ *is a PL 2-sphere and* $f(K^3) \subset \text{Int } h(\text{Bd } K^3)$.

PROOF. We pick $\delta < \varepsilon/2$ so that any δ-set in K^3 lies in a connected $\varepsilon/2$-set in K^3. Then $S^2 = h(\text{Bd}(K^3))$ of Theorem X.2.B. Let $D_1^2, D_2^2, \ldots, D_n^2$ be δ-disks on S^2 containing $S^2 \cap \text{Bd } K^3$. Obtain f by using Theorem X.2.A to pull K^3 through these disks into Int S^2. □

X.3. Repeated pull-backs. Repeated applications of Theorem X.2.C gives additional results.

THEOREM X.3.A. *Suppose* K^3 *is a crumpled cube in* R^3 *and* $\varepsilon > 0$. *Then there is an* ε-*homeomorphism* $h: K^3 \to R^3$ *such that for each* $\delta > 0$, *there is a* δ-*homeomorphism of* $h(\text{Bd } K^3)$ *into a PL 2-sphere in* $R^3 - h(K^3)$.

PROOF. A homeomorphism h such as promised by Theorem X.3.A can be described as the limit of a sequence of homeomorphisms $h_1, h_2 h_1, h_3 h_2 h_1, \ldots,$ where for some small ε_i, h_i is an ε_i-homeomorphism of $h_{i-1} \cdots h_1(K^3)$ as given by Theorem X.2.C. Also $h_i, h_{i-1} \cdots h_1(K^3)$ lies in the interior of a PL 2-sphere S_i such that there is an ε_i-homeomorphism of $h_{i-1} \cdots h_1(\text{Bd } K^3)$ onto S_i.

The ε's are chosen as follows. Let $\varepsilon_1 = \varepsilon/2$. Also, $\varepsilon_{i+1} < \varepsilon_i/2$. This insures that h is an ε-map. To insure that each $h_j \cdots h_1(K^3)$ and S_j lies in S_i for $j \geqslant i$, pick $\varepsilon_{i+1} < d(S_i, h_i \cdots h_1(K^3))/2$. To insure that h is one-to-one the ε_i's are restricted further. Let δ_i be a positive number so small that for $p, q \in K^3$ with $d(p, q) > 1/2^i$, then $d(h_i \cdots h_1(p), h_i \cdots h_1(q)) > \delta_i$. Pick $\varepsilon_{i+1} < \delta_i/2$. See Figure X.3.A. $\square$

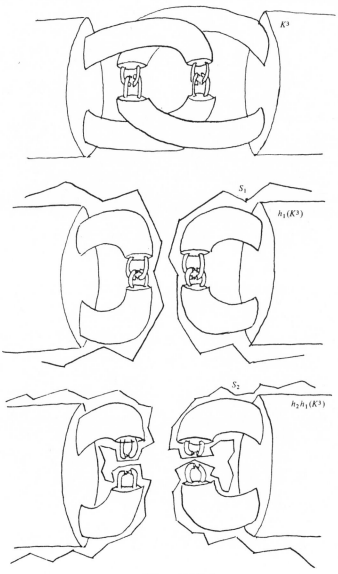

Figure X.3.A

The following is a restatement of Theorem X.3.A. Other versions of it are discussed in §6 of Chapter XVIII.

THEOREM X.3.B. *For each crumpled cube K^3 in R^3 and each $\varepsilon > 0$ there is an ε-homeomorphism h: $K^3 \to R^3$ such that $R^3 - h(K^3)$ is 1-ULC.*

PROOF. Let h be as given in Theorem X.3.A. To show that $R^3 - h(K^3)$ is 1-ULC suppose $\varepsilon' > 0$ is given and we are challenged to find a $\delta > 0$ such that any map f of Bd Δ^2 into a δ-subset of $R^3 - h(K^3)$ can be extended to map Δ^2 into an ε'-subset of $R^3 - h(K^3)$. Let δ be a positive number so small that each δ-subset of $h(\text{Bd } K^3)$ lies in a disk in $h(\text{Bd } K^3)$ of diameter less than $\varepsilon'/4$. Since the S_i's uniformly approach $h(\text{Bd } K^3)$, for i sufficiently large, δ-subsets of S_i lie in disks in S_i of diameter less than $\varepsilon'/2$. We show that such a δ meets the challenge.

Let f be a map Bd Δ^2 into a δ-subset of $R^3 - h(K^3)$ and g be an extension of f to a δ-set in R^3. Suppose that i is an integer so large that $f(\text{Bd } \Delta^2)$ lies in the unbounded component of $R^3 - S_i$ and there is a disk D_i in S_i of diameter less than $\varepsilon'/2$ that contains $S_i \cap g(\Delta^2)$. The Tietze extension theorem gives a map g' of Δ^2 into D_i such that g' agrees with g on $f'^{-1}(S_i)$. A suitable extension of f to Δ^2 is g on the component of $\Delta^2 - g^{-1}(S_1)$ containing Bd Δ^2 and g' elsewhere on Δ^2.

Note that instead of using an h on K^3 to make $R^3 - h(K^3)$ 1-ULC we could have used a homeomorphism h' on $R^3 - \text{Int } K^3$ to make $R^3 - h'(R^3 - \text{Int } K^3)$ 1-ULC.

THEOREM X.3.C. *Each 2-sphere in R^3 can be approximated by a 2-sphere each of whose complementary domains is 1-ULC.*

PROOF. A suitable approximation of Bd K^3 is $h_2 h_1(\text{Bd } K^3)$ where h_1 is a homeomorphism given by Theorem X.3.B to make $R^3 - h_1(K^3)$ 1-ULC and h_2 is a homeomorphism on $R^3 - h_1(\text{Int } K^3)$ to make $R^3 - h_2(R^3 - h_1(\text{Int } K^3))$ 1-ULC. $\square$

This line of attack on Theorem X.3.C is not recommended since it uses the side approximation theorem to get a weaker result.

Question X.3.D. Is there an easy procedure (say avoiding anything as hard as the side approximation theorem) for applying the techniques of Theorem X.1.A to arbitrary 2-spheres in R^3 to change them to 2-spheres with 1-ULC complements?

X.4. Sewing cubes together. As mentioned in Chapter IV, one gets a model for S^3 if two 3-cells are attached along their boundaries with a homeomorphism. We mentioned (but did not prove) that we also get a model of S^3 if two copies of Alexander horned spheres are attached with the identity homeomorphism along their boundaries. As pointed out in §5 of Chapter XVIII, one of the consequences of the side approximation theorem is that a crumpled cube W^3 is topologically a 3-ball if Int W^3 is 1-ULC. Theorem X.3.B has the following interesting application.

THEOREM X.4.A. *One obtains a model of S^3 by attaching an arbitrary crumpled cube to a real cube with a homeomorphism between their boundaries.*

PROOF. Consider $K^3 \cup_h W^3$ where K^3, W^3 are crumpled cubes, Int W^3 is 1-ULC, and h is a homeomorphism of Bd K^3 onto Bd W^3. It follows from Theorem X.3.C that there is an embedding r of K^3 into R^3 such that $S^3 - r(K^3)$ is 1-ULC. Then W^3 and $S^3 - \text{Int } r(K^3)$ are 3-balls and there are homeomorphisms f, g of W^3, $S^3 - \text{Int } r(K^3)$ respectively into Δ^3. Let ϕ be a homeomorphism of Δ^3 onto itself such that $f^{-1}\phi g = r$ on Bd W^3. A homeomorphism of $K^3 \cup_h W^3$ onto S^3 can be described as r on K^3 and $f^{-1}\phi g$ on W^3. $\square$

X.5. Pulling feelers off a sequence of disks. A *null sequence* of sets is one such that for each $\varepsilon > 0$, at most finitely many of the sets of the sequence have diameters more than ε. In this section we consider how we may pull-back feelers from a null sequence of disks. We generalize Theorems X.1.A and X.1.C as follows.

THEOREM X.5.A. *Suppose that*

X is a compact locally connected continuum,

$D_1, D_2, \ldots,$ is a null sequence of mutually disjoint tame disks such that each $X \cap \text{Bd } D_i = \varnothing$ and no Bd D_i links any simple closed curve in X,

$V_1, V_2, \ldots,$ is a sequence of disjoint open sets with Int $D_i \subset V_i$, and

M is a continuum in $X - \cup D_i$.

Then there is a map f of X into $(X \cup \cup V_i) - \cup D_i$ such that

f is fixed on M,

any point that is moved at all is moved into $\overline{\cup V_i}$,

if $p \in X - \cup V_i$ and p is moved by f, then some D_i separates p from M in X.

Furthermore, if ε is a positive number such that each V_i is of diameter less than ε and any point that D_i separates from M in X lies in the ε-neighborhood of D_i, then there is such a map f with $d(f, \text{Id}) \leq 2\varepsilon$.

PROOF. We suppose that V_i's are thin neighborhoods of the Int D_i's and form a null sequence. If this is not already the case, we reduce the V_i's to bring this about. As in the proof of Theorem X.1.C, we use Theorem X.A.1 to get a sequence of homeomorphisms $h_1, h_2 h_1, h_3 h_2 h_1, \ldots,$ of X into R^3. We use f_n to denote $h_n h_{n-1} \cdots h_1$ and define $f(p)$ to be $\lim f_n(p)$.

We now check that $f(p)$ is well defined for each $p \in X$. It is clearly well defined if there is an integer n_0 such that for each $j > n_0$, $f_j(p) = f_{n_0}(p)$. Let $n_1, n_2, \ldots,$ be the collection of all integers n_i such that $f_{n_i}(p) \neq f_{n_i-1}(p)$. Consider the case where this is an infinite increasing sequence. Then $f_{n_i}(p) \in V_{n_i}$. Let pq be an arc irreducible from p to M and p_i be the first point of Bd V_{n_i} on $f_{n_i}(pq)$ in order from $f_{n_i}(p)$ to q. Then $p_i \notin \cup V_j$ and $f_j(p_i) = p_i$ for $j \leq n_i$. Hence $p_{i+1} \in p_i q$ of pq. Hence $\lim p_i$ is a unique point p_0 of pq. Since $\{V_i\}$ is a null sequence, $\lim f_n(p)$ is this same point p_0 and $\lim f_n(p)$ is well defined.

We now consider the continuity of f at an arbitrary point $p \in X$. We show that f is continuous by showing that if $V(f(p), \varepsilon)$ is the open ε-neighborhood of $f(p)$ in R^3, then there is an open subset U_p of X containing p such that $f(U_p) \subset V(f(p), \varepsilon)$. The proof breaks into three cases.

Case 1. If for infinitely many integers j, $f_j(p) \neq f_{j-1}(p)$, we pick the monotone increasing sequence of integers $n_1, n_2, \ldots$, such that $f_{n_i}(p) \neq f_{n_i-1}(p)$ and let U_p be a connected open subset of X containing p such that $f_{n_1}(U_p) \subset V_{n_1}$. One finds that for each i, $f_{n_i}(U_p) \subset V_{n_i}$ and $f(U_p) = f(p) \subset V(f(p), \varepsilon)$.

Case 2. If for some n_0, $f_{n_0}(p) \in V_{n_0}$ and for $j \geq n_0$, $f_j(p) = f_{n_0}(p)$, we let U_p be a connected open subset of X containing p such that $f_{n_0}(U_p) \subset V(f(p), \varepsilon) \cap V_{n_0}$ and note that $f(U_p) = f_{n_0}(U_p) \subset V(f(p), \varepsilon)$.

Case 3. If for each j, $f_j(p) \in X - \cup V_i$, then each $f_i(p) = p$ and $f(p) = p$. Let n_0 be an integer so large that for $j \geq n_0$, diameter $V_{n_0} < \varepsilon/2$ and U_p be an open subset of X containing p such that for each point t of U_p there is an arc pt in X such that $f_{n_0}(pt)$ is of diameter less than $\varepsilon/2$. $\square$

One might notice that f would not be a homeomorphism if X contained a point p_0 such that infinitely many D_i^2's separate p_0 from M. We can put additional restrictions on X and M to insure that we get a homeomorphism.

THEOREM X.5.B. *Suppose*

X is a compact locally connected continuum with no separating points,

$D_1, D_2, \ldots$, is a null sequence of disjoint tame disks such that each $X \cap$ Bd $D_i = \varnothing$ and no Bd D_i links any simple closed curve in X,

$V_1, V_2, \ldots$, is a sequence of disjoint open sets with Int $D_i \subset V_i$, and

M is a nondegenerate continuum in $X - \cup D_i$.

Then there is a homeomorphism h of X into $(X \cup \cup V_i) - \cup D_i$ such that

h is fixed on M,

any point that is moved at all is moved into $\cup V_i$,

if $p \in X - \cup V_i$ and p is moved at all, then some D_i separates p from M in X.

Furthermore, if ε is a positive number such that each V_i is of diameter less than ε and any point that D_i separates from M in X lies in the ε-neighborhood of D_i, then there is such a homeomorphism h such that $d(h, \mathrm{Id}) \leq 2\varepsilon$.

PROOF. The required homeomorphism h is $\lim f_n$ as given in the proof of Theorem X.5.A. The reason that the limit is a homeomorphism rather than merely a map is that for each point p, there is an integer n_0 such that for $i > n_0$, $f_{i+1}(p) = f_i(p)$. To see that this is true, we would note from the proof of Theorem X.5.A that if for infinitely many n, $f_n(p) \neq f_{n-1}(p)$, then for each arc A from p to M, one has $\lim f_n(p) = h(p) \in A$. But there is an arc A' from p to M that misses $h(p)$ so it is impossible that for infinitely many n, $f_n(p) \neq f_{n-1}(p)$. Since no two points have the same image under any f_n, no two have the same image under h. $\square$

THEOREM X.5.C. *Theorem X.5.B holds if instead of supposing that X has no separating points and M is nondegenerate, we suppose instead that if $p \in X - M$, then not more than a finite number of D_i's separate p from M in X.* $\square$

In Chapter XI we give another result about pulling feelers of X off of a sequence (not necessarily a null sequence) of disks.

CHAPTER XI

INTERSECTIONS OF SPHERES AND 1-SIMPLEXES

We start this chapter with the elementary result that if K^2 is a 2-sphere in R^3 and $\varepsilon > 0$, then there is a triangulation T of R^3 of mesh less than ε such that $K^2 \cap T^0 = \emptyset$ and $K^2 \cap T^1$ contains no arc. Then we apply the pulling-back-feelers techniques of the last chapter to show that for each $\delta > 0$ there is a δ-homeomorphism $h: K^2 \to R^3$ such that $T^1 \cap h(K^2)$ is finite.

XI.1. Picking a triangulation. The proof of the following theorem depends more on the completeness of R^3 that on previous results.

THEOREM XI.1. *For each 2-sphere K^2 and each $\varepsilon > 0$ there is a triangulation T of R^3 such that*
$$K^2 \cap T^0 = \emptyset,$$
$K^2 \cap T^1$ contains no arc, and
mesh $T < \varepsilon$.

PROOF. Take a triangulation T_1 (of R^3) of mesh less than ε. Put round 3-balls B_i about the vertices v_i of T_1 such that if v_i' is a point in B_i, there is a homeomorphism h of R^3 onto itself such that h is linear on T, $h(v_i) = v_i'$, and the images of simplexes of T have diameters less than ε. Pick v_i' in Int $B_i - K^2$ and let h_1 be the linear homeomorphism.

Pick a ball B_i' about $v_i' = h_1(v_1)$ such that B_i' misses K^2 and pick $v_i'' \in B_i'$ such that the midjoint of no $v_i'' v_j''$ lies on K^2. Let h_2 be the linear homeomorphism that takes the v_i''s to the v_i'''s.

Continue refining the position of the v_i's so that the limit h has the property that for each pair of vertices v_i, v_j of T, neither the ends, the midpoints, the third way points, or any points that divide intervals from $h(v_i)$ to $h(v_j)$ in a rational ratio lie on K^2. The triangulation of R^3 whose simplexes are the image under h of the simplexes of T_1 satisfies the conclusion of Theorem XI.1. □

XI.2. Simplifying $K^2 \cap \Delta^1$. Suppose Δ^1 is a 1-simplex of a triangulation T selected in §XI.1. Then $K^2 \cap \mathrm{Bd}\, \Delta^1 = \varnothing$ and $K^2 \cap \Delta^1$ contains no arc. Let U be an open subset of R^3 containing $K^2 \cap \Delta^1$. We describe a homeomorphism $h: K^2 \to K^2 \cup U$ such that $\Delta^1 \cap h(K^2)$ is finite and h is the identity on $K^2 - U$. There are other proofs (see Bing [$\mathbf{B_{17}}, \mathbf{B_{18}}, \mathbf{B_{25}}$] and Craggs [$\mathbf{C_{11}}$]) that give even stronger results (such as the existence of such an h that extends to a homeomorphism of R^3) but we opt for this simpler proof since it is not dependent on more complicated results. An Appendix lists many standard geometric results that are used without proof in Chapter XI.

A set is *totally disconnected* if each of its components is a point. A subset of a 1-complex is totally disconnected if and only if it contains no arc.

THEOREM XI.2. *For each 2-sphere K^2, each 1-simplex Δ^1 in R^3, each neighborhood U of $K^2 \cap \Delta^1$, and each $\varepsilon > 0$, there is an ε-homeomorphism h of K^2 into $K^2 \cup \Delta^1$ such that $\Delta^1 \cap h(K^2)$ is finite and h is fixed outside U.*

PROOF. We start by getting a PL $\varepsilon/2$-homeomorphism f of R^3 into R^3 such that f is fixed outside U, $f(K^2) \cap \mathrm{Bd}\, \Delta^1 = \varnothing$, and $f(K^2) \cap \Delta^1$ is totally disconnected. Let T be a triangulation of R^3 of mesh less than $d((K^2 \cap \Delta^1), (R^3 - U))$. Just as we shifted vertices of T_1 in the proof of Theorem XI.1, we shift those vertices of T whose stars (under T) lie in U so that the resulting PL homeomorphism f^{-1} is close to the identity, is fixed outside U, and causes $K^2 \cap f^{-1}(\Delta^1)$ to be totally disconnected. Then f (the inverse of f^{-1}) is a PL $\varepsilon/2$-homeomorphism that is fixed outside U. Also $\Delta^1 \cap f(K^2)$ is totally disconnected and $\mathrm{Bd}\, \Delta^1 \cap f(K^2) = \varnothing$.

To simplify notation we suppose without loss of generality that f is the identity, $\Delta^1 \cap K^2$ is totally disconnected, $K^2 \cap \mathrm{Bd}\, \Delta^1 = \varnothing$ and each component of U has diameter less than $\varepsilon/2$. As long as we take care that h is the identity on $K^2 - U$ and takes K^2 into $K^2 \cup U$, there is no further need to check that h is an ε-map. It may be convenient to regard Δ^1 as vertical, but if we do, it is only for simplicity in visualization.

Let $H_1, H_2, \ldots, H_p$ be a finite collection of mutually disjoint disks on $K^2 \cap U$ such that $K^2 \cap \Delta^1 \subset \bigcup \mathrm{Int}\, H_i$. Let $M = K^2 - \bigcup \mathrm{Int}\, H_i$. Note that M is connected and misses Δ^1.

Let $C_1^3, C_2^3, \ldots, C_m^3$ be a finite collection of mutually disjoint right circular cylinders in $U - M$ with axes on Δ^1 such that the bases of the C_i^3's miss K^3, and the interiors of the C_i^3's cover $K^2 \cap \Delta^1$. We call the part of $\mathrm{Bd}\, C_i^3$ off the bases of C_i^3 the *side* of C_i^3.

It is a bit of a nuisance for $K^2 \cap \mathrm{Bd}\, C_i^3$ to have a component P such that $K^2 - P$ has more than two components. As noted in 1.K of the Appendix, there is not an uncountable family of mutually disjoint continua on K^3 each of which separates K^2 into more than two pieces. Since we can vary the radius of a cross section of C_i^3 in uncountably many ways, we suppose with no loss of generality that no component of $K^2 \cap \mathrm{Bd}\, C_i^3$ separates K^2 into more than two pieces. It follows from Theorem IX.3.B that separating a 2-sphere into more than two pieces is a topological property. Hence, no component of $K^2 \cap$ Bd C_i^3 separates Bd C_i^3 into more than two pieces.

An *irreducible separator* of K^2 is a subset X such that $K^2 - X$ is not connected but for each proper subset X' of X, $K^2 - X'$ is connected. It follows from 1.I of the Appendix that each irreducible separator is closed. A connected set is *unicoherent* if when it is expressed as the union of two relatively closed connected proper subsets C_1, C_2, then $C_1 \cap C_2$ is connected. It follows from the unicoherence of S^2 that any irreducible separator of K^2 is connected. If X is an irreducible separator of K^2 and M_1 is a component of $K^2 - X$, then $\overline{M}_1 = M_1 \cup X$. This does not necessarily hold if X is merely a separating continuum of K^2.

It follows from 3.F of the Appendix that if $\{X_\alpha\}$ is an uncountable collection of mutually disjoint continua in K^2 each separating K^2 then all but at most a countable number of elements of $\{X_\alpha\}$ are irreducible separators. Hence we suppose the radii of cross sections of the C_i^3's are selected so that each component of $K^2 \cap$ Bd C_i^3 that separates K^2 is an irreducible separator. (This step is not essential but it does enable us to use the $X_1, X_2, \ldots, X_n$ described in the next two sections to be the boundary components of the M_0 described there. We could have used $M_0 \cup (\cup X_i)$ instead of $\overline{M}_0$, but this would have been a notational complication with which to contend through three chapters. It seemed best to avoid this by choosing more restrictive X_α's.)

Some components of $K^2 \cap$ Bd C_i^3 may separate the two ends of $C_i^3 C_1^3$ from each other on Bd C_i^3. Each such continuum is the intersection of a sequence of annuli $A_1, A_2, \ldots$, on the side of C_i^3 such that $A_{j+1} \subset$ Int A_j. Another component X of $K^2 \cap$ Bd C_i^3 may fail to separate the ends of Bd C_i^3 from each other on Bd C_i^3 and if X^+ is the union of X and the component of Bd $C_i^3 - X$ missing the ends of Bd C_i^3, then X^+ is cellular on Bd C_i^3.

The homeomorphism h of our proof of Theorem XI.2 will result from modifying K^2 near the C_i^2's. The procedure is described in the next three sections. There may be some simplification in visualizing the components of $K^2 \cap C_i^3$ as annuli or disks, but a closer examination of the details of proof will show that no such restrictions are actually necessary. However, the reader is encouraged to get a preliminary uncomplicated view of the plan of attack before worrying about details.

XI.3. Shrinking at the waist. If C^3 is a cylinder and X is a continuum on the side of C^3, then X is called a *circling* (or *noncircling*) continuum of Bd C^3 if X separates (or does not separate) the bases of C^3 from each other on Bd C^3. In §XI.4 we discuss the elimination of noncircling components of $K^2 \cap$ Bd C_i', and in §XI.5 we discuss the elimination of circling components. In the present section we introduce techniques to be used in §§XI.4 and XI.5 by considering how we would proceed in the special case where there are no noncircling components of any $K^2 \cap$ Bd C_i^3.

A map f is *invariant* on a set X if $f(X) = X$. Note that this is less restrictive than the condition that f be fixed on X.

In this special case we let M_0 be the component of $K^2 - \bigcup \mathrm{Int}\, C_i^3$ containing M. Note that M_0 does not intersect Δ^1. Also, M_0 has only a finite number of boundary components $X_1, X_2, \ldots, X_n$, as noted in 4.B of the Appendix. Each of these X_j's is a component of $K^2 \cap (\bigcup \mathrm{Bd}\, C_i^3)$ which is an irreducible separator. Our plan for getting $h(K^2)$ is to discard $K^2 - \overline{M}_0$ and change $\overline{M}_0$ to a 2-sphere by shrinking the X_j's to points. Figure XI.3 shows how we shrink X_j by shrinking C_i^3 at the waist.

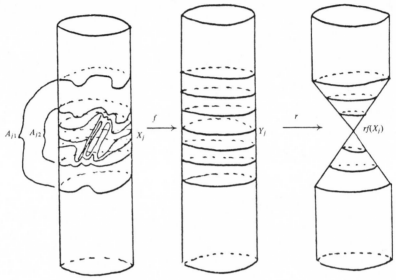

FIGURE XI.3

Let $A_{j1}, A_{j2}, \ldots,$ be a sequence of annuli on the sides of the C_i^3's such that $X_j = \bigcap_{k=1}^{\infty} A_{jk}$, $A_{j(k+1)} \subset \mathrm{Int}\, A_{jk}$, and $A_{j1} \cap A_{k1} = \varnothing$. Let f_1 be a homeomorphism of R^3 onto itself such that the $f_1(A_{j1})$'s are sections of the Bd C_i^3's determined by planes normal to Δ^1. Also, f_1 is invariant on the C_i^3's is the identity except near the sides of the C_i^3's (especially on M and outside U), and does not change a point's distance from Δ^1. Next we consider a homeomorphism f_2 of R^3 onto itself such that f_2 is invariant on the C_i^3's, agrees with f_1 except near the A_{j1}'s, and takes the A_{j2}'s onto normal sections of the Bd C_i^3's. We suppose $f_2(A_{j2})$ has the same horizontal midsection Y_j as $f_1(A_{j1})$ and its height is one half that of $f_1(A_{j1})$. Continuing in this fashion we get a sequence of f's and note that $\lim f_i = f$ determines a homeomorphism of $(R^3 - \bigcup X_j)$ onto $(R^3 - \bigcup Y_j)$ and is the identity on M and outside U and is invariant on $\bigcup C_i^3$. Perhaps f is not defined on $\bigcup X_j$ but let it be the set map $f(X_j) = Y_j$.

Let r be a map of R^3 onto itself suggested by the right part of Figure XI.3. The disk bounded by Y_j is taken to the center of this disk; r is a homeomorphism off such disks; points are either left fixed or moved toward Δ^1; $r(f(A_j^1))$ is the union of two cones with a common vertex and fixed bases; r is the identity on Δ^1 and outside U.

Although we cannot assume that f is a map on R^3 because it is not necessarily continuous on the X_j's, rf is a map since f sends points that are close to X_j close to

Y_j and r sends Y_j to a point. Also, rf sends $\overline{M}_0$ to a continuum homeomorphic to the decomposition space of $\overline{M}_0$ whose nondegenerate elements are the X_i's. This is the same as the decomposition of K^2 whose nondegenerate elements are the closures of components of $K^2 - M_0$. It follows from 5.B of the Appendex that such a decomposition yields a 2-sphere. In the special case where $K^2 \cap$ Bd C_i^3 contains only circling components, $rf(\overline{M}_0)$ is a 2-sphere in $K^2 \cup U$ and h is any homeomorphism of K^2 onto this 2-sphere that is fixed on M.

In §XI.5 we will speak of maps of the rf sort as ones which send certain X_j's to circles on the sides of the Bd C_i^3 and then shrink the circles to points. The difference in this future procedure is that we will use a map g (as described in §XI.4) to move a part of K^2 out of the way before performing the shrink r. Parts of $\overline{M}_0$ might lie in the Int C_i^3's and would need to be moved out before applying the shrink r.

XI.4. Removing noncircling components. Consider a C_i^3 and the components of $K^2 \cap$ Bd C_i^3. There may be uncountably many such components. Some may be noncircling components and some may separate the ends of C_i^3 on Bd C_i^3. It is our immediate goal to remove the noncircling components. In so doing we may inadvertently remove some circling ones. We do not mind. Before removing any components, let us see how the components lie on Bd C_i^3.

The circling components of $K^2 \cap$ Bd C_i^3 are ordered on Bd C_i^3 and have a closed union W as noted in 4.C of the Appendix. We suppose $W \neq \varnothing$. A component of Bd $C_i^3 - W$ is either an open annulus or an open disk containing an end of C_i^3. Let $D_{i1}, D_{i2}, \ldots$, be a countable number of mutually disjoint disks in Bd C_i^3 such that the D_{ik}'s miss W and the ends of C_i^3, $(K^2 \cap$ Bd $C_i^3) - W \subset \bigcup_{k=1}$ Int D_{ik}, and no compact set in Bd $C_i^3 - W$ intersects more than a finite number of D_{ik}'s. Consider such disks D_{ik} from each Bd C_i^3 and let $D_1, D_2, \ldots$, be an ordering of all the disks from all the Bd C_i^3's.

We are getting ready to apply Theorem X.1.B so we raise small blisters on each side of each D_k with E_k being the top of the blister in the appropriate C_i^3 and F_k the top of the outside blister. Let B_k^3 be the union of the two blisters with a common base D_k. It is a tame 3-ball. We suppose that the B_k^3's are mutually disjoint, and each point of each B_i^3 is close to D_k—say within $1/k$ of J.

Let D_1, K^2, M, B_1^3 be the D^2, X, M, B^3 of the statement of Theorem X.1.B and g_1 be the promised homeomorphism. Note that $g_1(K^2) \subset (K^2 \cup$ Int $B_i^3) - D_i^2$. It resulted from pulling back feelers that went through Int B_i^3. In fact, $g_1(K^2)$ is the union of a subset of B_1^3 and all components V of $K^2 - B_1^3$ such that any arc from M to V crosses D_1^2 algebraically a nonzero number of times. Note that g_1 is a homeomorphism on K^2 rather than on R^3.

Skip the description of g_2 if $g_1(K^2) \cap D_2 = \varnothing$ and let $g_2 = g_1$. Also, we do not consider B_2^3. By disregarding such B^3's we may make our final sequence of B^3's locally finite at more points. A collection G is *locally finite at a point* if there is a neighborhood of p that intersects at most a finite number of elements of G.

In the more likely case that $g(K^2) \cap D_2 \neq \varnothing$, we let $D_2, g_1(K^2), M, B_2^3$ be the D_2, X, M, B^3 of the statement of Theorem X.1.B and ϕ be the promised homeomorphism. Denote ϕg_1 by g_2. Note that

$$g_2(K^2) \subset \left(g_1(K^2) \cup \operatorname{Int} B_2^3\right) - D_2 \subset K^2 \cup \operatorname{Int} B_1^3 \cup \operatorname{Int} B_2^3 - (D_1 \cup D_2).$$

The operation is repeated infinitely often. Denote $\lim g_i$ by g. We do not claim that g is continuous on certain circling components of $K^2 \cap (\cup \operatorname{Bd} C_i^3)$ on which each g_i is fixed. It is to be noted that if g_j leaves some point of some component of $K^2 \cap (\cup \operatorname{Bd} C_i^3)$ fixed, it leaves the whole component fixed. When g_j sends a point off $\cup \operatorname{Bd} C_i^3$, all succeeding g_k's do also.

Let $\{X_\alpha\}$ be the collection of all components of $g(K^2) \cap (\cup \operatorname{Bd} C_i^3)$. These components are circling components that are left fixed by each g_i. It follows from 4.C of the Appendix that $\cup X_\alpha$ is closed and from 4.B of the Appendix that the boundary of the component M_0 of $K^2 - \cup X_\alpha$ containing M is the union of a finite number of X_α's—say $X_1, X_2, \ldots, X_n$. It can be shown that the selection of the $X_1, X_2, \ldots, X_n$ is affected by the ordering of the D_k's, but we shall not pursue that.

We now show that g is a homeomorphism on M_0. Consider a point $p \in M_0$, and a compact connected set Z in M_0 such that $M \subset Z$ and some neighborhood of p in K^2 lies in Z. We show that g is a homeomorphism at p by showing that for some large integer n, $g_i = g$ on Z if $i > n$. For each $p \in Z$ there is an integer n_1 such that $g_{n_1}(p) \notin \cup \operatorname{Bd} C_i^3$. Then there is a neighborhood N of p in K^2 such that if $n \geqslant n_1$, $g_n(N) \cap (\cup \operatorname{Bd} C_i^3) = \varnothing$. A finite number of such N's cover Z so there is an integer n_2 such that if $n > n_2$, $g_n(Z) \cap \cup \operatorname{Bd} C_i^3 = \varnothing$.

Now $g_{n_2}(Z)$ misses most B_k^3's so that there is an n_3 such that if $n \geqslant n_3$, $g_{n_2}(Z) \cap \cup_{n_3}^\infty B_k^3 = \varnothing$. Then $g = g_{n_3}$ on Z and g is a homeomorphism at $p \in M_0$.

Henceforth in the proof of Theorem XI.2 we shall use $\overline{M}_0$ instead of K^2. If the X_i's are shrunk to points, $\overline{M}_0$ becomes a 2-sphere. The map g removes the noncircling components of $\overline{M}_0 \cap \cup \operatorname{Bd} C_i^3$. It may inadvertently remove some circling components of $\overline{M}_0 \cap \cup \operatorname{Bd} C_i^3$ if these lie on feelers that poke through D_j's.

It might be useful to summarize how $g(M_0)$ was obtained. The open subset M_0 of K^2 has the following properties.

$M \subset M_0$,

M_0 has only a finite number of boundary components $X_1, X_2, \ldots, X_n$ and each is a circling component of some C_i^3,

the g_i's associated with the D_k's do not move the X_j's and each component of $M_0 \cap \cup \operatorname{Bd} C_i^3$ is moved off of $\cup \operatorname{Bd} C_i^3$ by some g_j.

If A is a continuum in $\overline{M}_0$ that intersects M but misses the B_k^3's, both g and g^{-1} are the identity on A.

Successively, part of K^2 is sucked into open 3-balls spanned by the D_k's. This is done so that the resulting 2-sphere does not intersect $D_1 \cup D_2 \cup \cdots \cup D_k$. Note that g is not defined on R^3 but only on K^2. Note that g is the identity on M. We shall only be concerned with g defined on $\overline{M}_0$. $\square$

For future reference we need to know where $\{B_j\}$ is locally finite on $\overline{M}_0 \cap \operatorname{Bd} C_i^3$. First we note that it need not be locally finite at points of the boundary of

M_0 since there may be D_k^2's converging to X_n's. If p lies on a noncircling component of $K^2 \cap \mathrm{Bd}\, C_i^3$, it lies on the interior of a single D_r^2 so $\{B_k^3\}$ is locally finite there. If $p \in M_0$ belongs to a circling component of $K^2 \cap \mathrm{Bd}\, C_i^3$, this component is not an X_r but at some stage was pulled into some B_s^3. Also, D_t^2's near p would also be pulled inside this same B_s^3. The B_t^3's corresponding to these D_t^2 would not have been included in our list of B^3's and this makes the acceptable list of B^3's locally finite at each point of M_0.

XI.5 Removing circling continua. After we have obtained an open subset M_0 of K^2 with only a finite number of boundary components and obtained a homeomorphism g of M_0 into $R^3 - \cup C_i^3$ we are not in quite as simple a situation as we were in the special case of §XI.3. The difference is that g may not be continuous on the boundary of M_0. However, this offers no real problem since we are going to shrink the boundary components of M_0 to points with a map rf as given in §XI.3, and rfg is continuous on $\overline{M}_0$. Recall that $f_{i+1} = f_i$ except near A_i so f is continuous on $R^3 - \cup X_i$.

We do not define h to be rfg—rather h is any homeomorphism whatsoever of K^2 onto $rfg(\overline{M}_0)$ as long as h is the identity on M. If B_j^+ denotes a typical outer blister $B_k^3 - C_i^3$ (where C_i^3 intersects B_k^3), then $g(M_0) \subset (M_0 - \cup C_i^3) \cup (\cup B_j^+)$. An interesting observation is that if A is a continuum in M_0 intersecting M but missing $\cup F$, then g and g^{-1} are the identity on A. Recall that the F's were part of the boundaries of the B^+'s. $\square$

In order to simplify the proof of Theorem XI.2 we did not modify K^2 near the 1-skeleton T^1 of triangulation T of Theorem XI.1 but only modified it near Δ^1. However, we could have covered $K^2 \cap T^1$ with C_i^3's and have gotten the following.

THEOREM XI.5. *For each 2-sphere K^2, each triangulation T of R^3, each neighborhood of $K^2 \cap T^1$, and each $\varepsilon > 0$ there is an ε-homeomorphism h of K^2 into $K^2 \cup U$ such that $T^2 \cap h(K^2)$ is finite and h is fixed outside U.*

Henceforth, when we speak of r, f, g, M_0 we will be considering modifying K^2 near T^1 rather than near Δ^1. $\square$

In Theorem XI.6 we shall need to give special attention to the B_k^3's. Recall that $\{B_k^3\}$ is locally finite at each point of M_0. Also, $\{rf(B_k^3)\}$ is locally finite except at points of $rf(\mathrm{Bd}\, M_0)$. Hence they constitute a null sequence. We suppose that the f_k's were selected so that $\{rf(B_k^3)\}$ is a null sequence of round balls missing T^2 as shown in Figure XI.5. We shall make use of the fact that $T^2 \cap rf(\cup B_k^3) = \varnothing$.

XI.6. Locating $h(K^2)$. In Chapters XII and XIII we shall refer to the $h(K^2)$ we have described in the present chapter. We shall give special attention to certain properties of $h(K^2)$. We end the present chapter by giving certain properties of $h(K^2)$ that will be used later.

Just where is $h(K^2) = rfg(\overline{M}_0)$? Since $g(M_0) \subset R^3 - \cup C_i^3$ and

$$g(M_0) \subset \left(M_0 - \cup B_k^3\right) \cup \cup \left(B_k^3 - D_i\right),$$

$$g(M_0) \subset \left(M_0 - \cup C_i^3 - \cup B_k^3\right) \cup \left(\cup B_k^+\right).$$

Also,

$$rfg(M_0) \subset \left(rf(M_0) - \cup rC_i^3 - \cup rf(B_k^3)\right) \cup \left(rf(B_k^+)\right).$$

We may obtain $h(K^2) = rfg(\overline{M}_0)$ by starting with $rf(\overline{M}_0)$, removing pinched cylinders (the rC_i^3's), putting back the pinch points, removing the union of a null sequence of 3-balls ($rf(B_k^3)$'s), throwing away certain components of the remainder but keeping others, and then adjoining to the kept part a subset of the union of the $rf(B_j^+)$'s. Figure XI.5 shows some $rf(B_j^3)$'s.

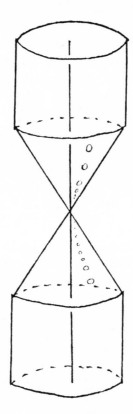

FIGURE XI.5

If we had taken C_i^3 to be PL rather than circular we could have made the B_k^3's PL. A map F is called PL mod $\{X_\alpha\}$ if for each $p \in$ (domain of F) $- \cup X_\alpha$, F is locally PL at p. We could have made the f's PL mod the X_j's and r PL mod the $r^{-1}r(Y_j)$'s so we suppose this. This causes the $rf(B_k^3)$'s to constitute a null sequence of PL 3-balls such that no compact set in $R^3 - \Delta^1$ intersects more than a finite number of $rf(B_k^3)$'s. With a homeomorphism of R^3 onto itself we can shrink each of these PL 3-balls $rf(B_k^3)$ to a very small PL 3-ball near any interior point of $rf(B_i^3)$ so we suppose without loss of generality that if T is the triangulation of Theorem XI.1, then each $rf(B_k^3)$ misses T^2. The fact that the $rf(B_k^3)$'s miss T^2 is very important for Theorem XI.6, Theorem XII.5, and Chapter XIII. We mention some properties of $S^2 = rfg(\overline{M}_0)$.

(1) Mesh of T is small.

(2) $T^0 \cap K^2 = \varnothing$.

(3) $T^1 \cap K^2$ is totally disconnected.

(4) The C_i^3's are small.

(5) $S^2 = h(K^2) = rfg(\overline{M}_0)$ is a 2-sphere.

(6) $T^0 \cap S^2 = \varnothing$.

(7) $T^1 \cap S^2$ is finite.

(8) rfg is near the identity on $\overline{M}_0$.

(9) h is near the identity on K^2.

(10) Each component of $K^2 - M_0$ is small.

(11) T^1 pierces S^2 at each point of $T^1 \cap S^2$.

(12) $rf: R^3 \to R^3$ is the identity except near the C_i^3's.

(13) rf is near the identity. (This is not proven until Chapter XII.)

(14) f is PL mod the X_j's.

(15) The f_i's are PL on R^3.

(16) r is PL mod the $r^{-1}r(Y_j)$'s.

(17) r is the limit of r_i's such that each r_i is PL on R^3.

(18) Each $rf(B_k^3)$ misses T^2.

(19) $S^2 \subset (S^2 \cap T^1) \cup rf(K^2 - \cup C_i^3) \cup (rf(\cup B_k^+))$.

In Chapter XII we shall use the following observation about the $rfg(\overline{M}_0) = h(K^2)$ that we have described.

THEOREM XI.6. *Suppose in the description of S^2 that*

(1) *each component of $K^2 - M_0$ has diameter less than ε.*

(2) $d(rfg, \mathrm{Id}) < \varepsilon$ *on* $\overline{M}_0$,

(3) *each C_i^3 is of diameter less than ε, and*

(4) *Q is a closed subset of $T^2 \cap rfg(\overline{M}_0) = T^2 \cap S^2$ such that each componnent of $S^2 - Q$ has diameter less than ε.*

Then for each open set U in R^3 containing Q there is a PL ε-homeomorphism $H: R^3 \to R^3$ such that each component of $H(K^2) - U$ has diameter less than 7ε.

PROOF. While f need not be a homeomorphism, it is the limit of a sequence of PL homeomorphisms $f_1, f_2, \ldots$. Regard r also as the limit of such a sequence of PL homeomorphisms $r_1, r_2, \ldots$, such that for each neighborhood N of $T^1 \cap S^2$ there is a positive integer n_0 such that for $m > n_0$, $rf(r_m f_m)^{-1}$ is the identity on $R^3 - N$. Figure XI.6 shows $r_m f_m$ near a C_i^3. Since the waist of C_i^3 has been shrunk to a neck instead of to a point, there is room for a feeler to run through the neck without creating a singularity in the feeler.

Some $r_i f_i$ is an H satisfying the requirements of the theorem. The facts that the C_i^3's are small and the $r_i f_i$'s are the identity except near the C_i^3's causes $d(r_i f_i, \mathrm{Id})$ to be less than ε for all large i.

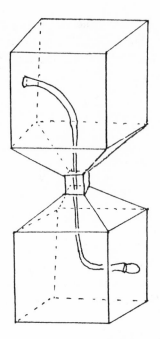

FIGURE XI.6

To finish the proof of the theorem we assume that it is false, U is an open subset of R^3 containing Q, and for each positive integer i, W_i is a component of $r_i f_i(K^2) - U$ with diameter as much as 7ε. Denote $(r_i f_i)^{-1}(W_i)$ by W_i' and note that diameter $W_i' > 5\varepsilon$.

Although $W_i' \subset K^2$, we do not know that it lies in $\overline{M}_0$. To handle this, we change W_i'. If W_i' intersects a component of $K^2 - \overline{M}_0$, we subtract the component from W_i' and add in the boundary of this component of $K^2 - \overline{M}_0$. Denote by W_i'' the continuum in $\overline{M}_0$ obtained by successively performing this replacement for each component of $K^2 - \overline{M}_0$ that W_i' touches. Then diameter $W_i'' > 3\varepsilon$.

Let W_∞ be a continuum in $\overline{M}_0$ which is the limit of a subsequence of $W_1'', W_2'', \ldots$. For convenience in notation we suppose that $W_1'', W_2'', \ldots$, *converges* to W_∞—if $p \in W_\infty$, each neighborhood of p intersects all but at most a finite number of W_i'''s, but if $p \notin W_\infty$ some neighborhood of p misses all but at most a finite number of W_i'''s. Note that the diameter of $W_\infty > 3\varepsilon$ and the diameter of $rfg(W_\infty) \geq \varepsilon$. We now show that $rfg(W_\infty) \subset S^2 - Q$. Consider a point $p \in W_\infty$. If none of the g_i's move p, $rfg(p) = \lim r_i f_i(p)$. Since none of the $r_i f_i(p)$ lie in U, $rf(p)$ does not lie in U and is not a point of Q. If p is moved by some g_i, it is moved by only a finite number of g_i's. The last of these moves p into a B_i^3 and the $rf(B_i^3)$'s miss T^2. Hence $rfg(p) \notin Q$. The facts that $rfg(W_\infty) \subset K^2 - Q$ and diameter $rfg(W_\infty) > \varepsilon$ violates the hypotheses of Theorem XI.6. The assumption is false that each $r_i f_i(K^2) - U$ contains a big W_i. $\square$

INTERSECTIONS OF SURFACES WITH SKELETONS

Suppose K^2 is a topological 2-sphere (perhaps wild) in R^3 and T is a triangulation of R^3 with i-skeleton T^i ($i = 0, 1, 2$). Chapter XI gives a procedure for changing K^2 to S^2 so that $S^2 \cap T^0 = \varnothing$ and $S^2 \cap T^1$ is finite.

In Chapter XII we consider procedures for simplifying $S^2 \cap T^2$. We use Δ^2 to denote a 2-simplex—perhaps an element of T but this is not always required.

Three useful ideas are exploited in this chapter. (1) It may be helpful in studying a set to expand the set and examine the expansion. (2) Pushing at the center of a membrane may eliminate triods. (3) Spinning about an axis may improve intersections. We develop these three ideas in the first four sections of this chapter and then apply them in the fifth section to simplify $S^2 \cap T^2$. In the fifth section we apply the techniques we have developed to $h(K^2) = rfg(\overline{M_0}) = S^2$ of Chapter XI. The three ideas would seem to be of interest not only for their applications in Chapters XII and XIII but also because the information they provide about surfaces in R^3.

XIII.1. Expanding closed sets. Suppose X, Y are compact closed sets in R^3 (or Δ^2 or S^2) and $\{G_i\}$ is a null sequence of 3-balls (or disks) in $R^3 - Y$ (or $\Delta^2 - Y$ or $S^2 - Y$) such that $\{\text{Int } G_i\}$ covers $X - Y$ and each Int G_i intersects X. Then $X \cup (\cup G_i)$ is called *an expansion of X in $R^3 - Y$* (or $\Delta^2 - Y$ or $S^2 - Y$). If X is a complicated continuum, one of the methods used to investigate the properties of X is to examine expansions of it. Some expansions are locally connected, and locally connected continua may be easier to study than ordinary continua.

There are many expansions of X in $R^3 - Y$ (or $\Delta^2 - Y$ or $S^2 - Y$), but they may share certain properties. In applying expansions we shall frequently use $\Delta^2 \cap S^2$ as X and Bd Δ^2 as Y. Then X lies in each of R^3, Δ^2, and S^2 has expansions in each of $R^3 - Y$, $\Delta^2 - Y$, and $S^2 - Y$.

THEOREM XII.1.A. *If a closed set X lies in a 2-simplex Δ^2 such that $X \cap$ Bd Δ^2 contains no arc, each component of each expansion of X in Int Δ^2 is locally connected.*

PROOF. An expansion is locally connected except possibly at points of $X \cap$ Bd Δ^2. However, if a compact continuum is not locally connected, it contains a nondegenerate continuum where it is not locally connected. See 1.L of the Appendix. However, each component of $X \cap$ Bd Δ^2 is a one-point set since it contains no arc. Hence, any component of any expansion of X in Δ^2 with respect to $X \cap$ Int X is locally connected. $\square$

A set is *nondegenerate* if it contains more than one element and *degenerate* if it has only one. A set all of whose components are degenerate is called *totally disconnected*. Note that a subset of an arc is totally disconnected if and only if it contains no arc.

We obtain the following by an argument like that used to get Theorem XII.1.A.

THEOREM XII.1.B. *If X, Y are closed compact subsets of R^3 (or S^2) and $X \cap Y$ is totally disconnected, then each component of each expansion of X in $R^3 - Y$ (or $S^2 - Y$) is locally connected.* $\square$

Note that if $X = S^2 \cap \Delta^2$, X might be expanded into either $S^2 - Y$, $\Delta^2 - Y$, or $R^3 - Y$. However, the existence of an expansion of a certain sort in one might imply an expansion of a similar sort in the others.

A *triod* is the union of three arcs $a_1 p$, $a_2 p$, $a_3 p$ such that $a_i p \cap a_j p = \{p\}$ if $i \neq j$. Suppose X is a closed set in R^3 (or K^2 or Δ^2). An expansion $E(X)$ of X in $R^3 -$ Bd Δ^2 (or $K^2 -$ Bd Δ^2 or $\Delta^2 -$ Bd Δ^2) has the *notriod property* if there is no triod in $E(X)$ with ends on Bd Δ^2 and which lies except for these ends in $E(X) -$ Bd Δ^2.

In the following we shall use S^2 to denote a 2-sphere in R^3 (usually the image of another 2-sphere K^2—but that does not matter here) and Δ^2 to denote a 2-simplex (usually an element of T—but that does not matter here either).

THEOREM XII.1.C. *If there is an expansion of $S^2 \cap \Delta^2$ in one of $R^3 -$ Bd Δ^2, $S^2 -$ Bd Δ^2, or Int Δ^2 that has the no-triod property, then there is an expansion in each of the other two that has the no-triod property.*

PROOF. Suppose $E(R^3)$ is an expansion of $S^2 \cap \Delta^2$ in $R^3 -$ Bd Δ^2, and $E(R^3)$ has the no-triod property. Then $E(R^3)$ contains expansions of $S^2 \cap$ Int Δ^2 in $S^2 -$ Bd Δ^2 and $\Delta^2 -$ Bd Δ^2, respectively, and each of these expansions would have the no-triod property.

If $E(S^2)$ is an expansion of $S^2 \cap \Delta^2$ in $S^2 -$ Bd Δ^2 such that $E(S^2)$ as the no-triod property, we explain how to get an expansion $E(R^3)$ of $S^2 \cap \Delta^2$ in $R^3 -$ Bd Δ^2 with the no-triod property. Suppose $\{D_i\}$ is a collection of disks in $S^2 -$ Bd Δ^2 used in building $E(S^2)$. Let $\{B_j\}$ be a null sequence of round 3-balls in $R^3 -$ Bd Δ^2 used to build an $E(R^3)$ such that each $S^2 \cap B_j$ lies in some D_k. Then $E(R^3)$ contains no triod, because if it did, the D_i's could be used to find a triod in $E(S^2)$. Hence $E(R^3)$ has the no-triod property if $E(S^2)$ does.

Similarly, if some expansion of $S^2 \cap \Delta^2$ in $\Delta^2 -$ Bd Δ^2 has the no-triod property, then there is a similar expansion of $S^2 \cap \Delta^2$ in $R^3 -$ Bd Δ^2. $\square$

If X is a closed set and each component of $X \cap \text{Bd } \Delta^2$ is a point, we say that X *has the no-triod* property if there is an expansion $E(X)$ of X in $R^3 - \text{Bd } \Delta^2$ such that $E(X)$ has the no-triod property. For example, the $S^2 \cap \Delta^2$ shown in the left part of Figure XII.1.C has the no-triod property but the right does not.

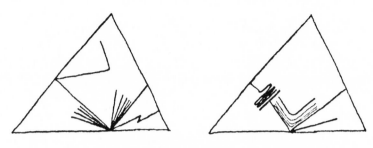

FIGURE XII.1.C

XII.2. Pushing at a membrane. Consider two tetrahedra Δ_1^3, Δ_2^3 which intersect each other at a common base Δ^2 and lie except for the base on opposite sides of it. Now consider a homeomorphism α_i^{-1} of R^3 onto itself which is fixed outside $\Delta_1^3 \cup \Delta_2^3$ and is conical in taking the join of $\text{Bd } \Delta_1^3 \cup \text{Bd } \Delta_2^3 - \text{Int } \Delta^2$ with the barycenter p_0 of Δ^2 linearly onto the join of $\text{Bd } \Delta_i^3 \cup \text{Bd } \Delta_2^3 - \text{Int } \Delta^2$ with a point p_i near p_0 but off Δ^2. We suppose that p_i's $(i = 0, 1, 2, \ldots)$ are chosen so that the intersection of any two such joins is $\text{Bd } \Delta^2$. The homeomorphisms acting on Δ^2 resemble pushes at the center of a drum Δ^2. We call each α_i^{-1} a *center push*. If α_i^{-1} is an ε-homeomorphism, we call it an *ε-center push*. If S^2 is a topological 2-sphere we shall seek an α_i such that $\Delta^2 \cap \alpha_i(S^2)$ is nicer than $\Delta^2 \cap S^2$.

THEOREM XII.2.A. *Suppose $S^2 \cap \text{Bd } \Delta^2$ is finite and α_i $(i = 1, 2, \ldots)$ are the homeomorphisms described in the preceding paragraph. Then some one of $\Delta^2 \cap \alpha_i(S^2)$ has the no-triod property.*

PROOF. Let n be an integer more than twice the number of unordered triples of points in $S^2 \cap \text{Bd } \Delta^2$. Let $E_1, E_2, \ldots, E_n$ be expansions of $S^2 \cap \alpha_i^{-1}\Delta^2$ in S^2 with respect to $S^2 - \text{Bd } \Delta^2$ such that no two of these expansions intersected each other except on $S^2 \cap \text{Bd } \Delta^2$. Assume the theorem is false, and Y_i is a triod in E_i with ends on $\text{Bd } \Delta_i^2$ and remainder off $\text{Bd } \Delta_i^2$. No three Y_i's have the same ends, or else S^2 would contain a skew curve and violate Theorem III.3.D. Hence some E_i has the no-triod property, and the associated $\Delta^2 \cap \alpha_i(S^2)$ has the no-triod property. $\square$

For simplicity we used PL α_i's in Theorem XII.2.A, and $\alpha_i^{-1}(\Delta^2)$ was the cone over a triangle. Had we wanted to keep α_i fixed on a closed set missing $\text{Int } \Delta^2$, and if these closed sets had come in tangential to Δ^2 at $\text{Bd } \Delta^2$, we could have picked $\alpha_i^{-1}(\Delta^2)$ to resemble a saggy tent. See Figure XII.2.A.

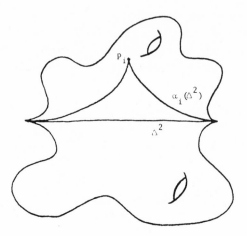

FIGURE XII.2.A
"Saggy Tent"

THEOREM XII.2.B. *Suppose* $S^2 \cap$ Bd Δ^2 *is finite and* U *is an open set in* R^3 *containing* Int Δ^2. *Then there is a homeomorphism* α *of* R^3 *onto itself such that*
  α *is the identity outside* U,
  each point that α *moves is moved in a direction normal to* Δ^2,
  α *is* PL mod $S^2 \cap$ Bd Δ^2,
  $\alpha^{-1}(S^2) \cap \Delta^2$ *has the no-triod property.* $\square$

Although Theorem XII.2.A and Theorem XII.2.B did not speak of $rfg(\overline{M_0})$ nor $rf(B_i^3)$'s, if $S^2 = rfg(\overline{M_0})$, we could have picked α to be the identity on the $rf_2(B_i^3)$'s.

XII.3. Eliminating nonpiercing points. Suppose S^2 is a 2-sphere and apb is an arc that intersects S^2 only at p. We say that apb *pierces* S^2 at p if a and b lie in different components of $R^3 - S^2$. If apb lies on a simple closed curve, line, or bigger arc, we say that the curve, line, or bigger arc *pierces* S^2 at p if apb does, irrespective as to how it intersects S^2 elsehwere. An arc apb *pierces a disk* D at a *point* p if $apb \cap D = \{p\}$ and there is an $\varepsilon > 0$ such that each arc from ap to bp in $R^3 - D$ is of diameter more than ε. We say that $ap - \{p\}$ and $bp - \{p\}$ *are on opposite sides of* D. It follows from methods of Chapters VIII and IX that apb pierces D if and only if apb lies on a simple closed curve that misses $D - \{p\}$ and links Bd D. If apb lies in a simple closed curve J, line L, or arc $a'pb'$, we say that J, L, or $a'pb$, *pierces* D at p if apb does (irrespective of how $J - apb$, $L - apb$, $a'pb' - apb$ intersects D). The following theorem says that S^2 can be modified to remove certain nonpiercing points.

THEOREM XII.3.A. *Suppose* S^2 *is a 2-sphere and* apb *is a straight arc such that* $S^2 \cap apb = \{p\}$, *but* apb *does not pierce* S^2 *at* p. *Then for each neighborhood* N *of* p *there is a* PL *homeomorphism* β: $R^3 \to R^3$ *such that*
  β *is the identity off* N *and*
  $\beta(S^2) \cap apb = \varnothing$.

PROOF. Let axb be a PL arc from a to b in $R^3 - S^2$ such that $axb \cup apb$ is a simple closed curve. Let P^3 be a triangular cylinder with apb as axis is shown in Figure XII.3.A such that $p \in \text{Int } P^3$, $P^3 \subset N$, S^2 does not intersect either base of P^3, and $P^3 \cap (axb \cup apb)$ is a subinterval $a'pb'$ of abp.

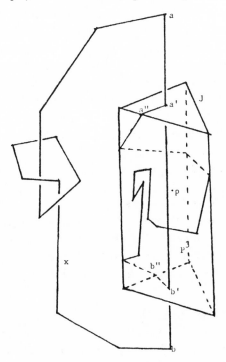

FIGURE XII.3.A.

Since each component of $R^3 - S^2$ is 0-ULC (see Theorem IX.3.D), a slight change can be made in arc apb near p to get a PL arc aqb which lies in one component of $R^3 - S^2$. Then $axb \cup aqb$ links J of Figure XII.3.A. It follows from Theorems VIII.1.C and VIII.5.B that $axb \cup aqb$ bounds a PL 2-manifold M^2 in $R^3 - S^2$. We suppose it is GP with respect to Bd P^3.

What does $M^2 \cap$ Bd P^3 look like? It is a 1-manifold-with-boundary with only two boundary points a', b'. Let $a'yb'$ be the arc from a' to b' in $M^2 \cap$ Bd P^3. The required PL homeomorphism β is any one fixed except near P^3 that sends $aa'yb'b$ to apb. Since $aa'yb'b$ misses S^2, $apb = \beta(aa'yb'b)$ misses $\beta(S^2)$. $\square$

A slightly different proof of this result is given in Theorem 3 of $[\mathbf{B_{16}}]$.

We now show that if S were obtained from K^2 as discussed in Chapter X where $S^2 = rfg(\overline{M_0})$, then L pierces S^2 at each point of $S^2 \cap L$, and there is no need to apply Theorem XII.3.A to S^2. Although we use Theorem XI.3.B in Chapter XII, its use there is not essential and could have been avoided. It is included since it throws light on the behavior of certain modifications of 2-spheres in R^3.

THEOREM XII.3.B. *Suppose* K^2, Δ^1, r, f, g, M_0 *are as in Theorem* XI.2 *and* $h(K^2) = rfg(\overline{M_0})$ *is the promised homeomorphism. Then if* $p \in \Delta^1 \cap h(K^2)$, Δ^1 *pierces* $h(K^2)$ *at* p.

PROOF. We show that if $p_i = rf(X_i)$ is a point of $h(K^2) = rfg(\overline{M_0})$, then Δ^1 pierces $h(K^2)$ at p_i by showing that for each neighborhood N of p_i in R^3 there is a simple closed curve in $N \cap h(K^2)$ that links the straight line L containing Δ^1.

Let A_1 be an annulus in K^2 such that $X_i \subset \text{Int } A_1$, $rfg(A_1 \cap \overline{M_0}) \subset N$, and each $L \cap g_j(A_1 \cap \overline{M_0}) = \varnothing$. Let J_1 be the boundary component of A_1 in M_0. We shall show that $rfg(J_1)$ links L.

Since A_1 is an ANR, there is an open set V in R^3 containing A_1 and a retraction ρ_1 of V onto A_1 that does not move any point as much as $d(L, A_1)$. Let A_2 be an annulus in the intersection of V and the side of the Bd C_j containing X_i. Let J_2 be a boundary component of A_2. Let ρ_2 be a projection of A_1 onto J_1.

Consider J_2, $\rho_1(J_2)$, $\rho_2\rho_1(J_2)$, and $rfg(J_1) = rfg\rho_2\rho_1(J_2)$.

Then J_2 crosses a half plane π bounded by L an odd number of times. Although $\rho_1(J_2)$ is perhaps singular, it is homotopic to J_2 in $R^3 - L$ and crosses π an odd number of times. Although $\rho_2\rho_1$ is a singular map, it sends J_2 to J_1, so J_1 crosses π an odd number of times since $\rho_2\rho_1(J_2)$ does.

To see that $rfg(J_1)$ links L we only show that $g(J_1)$ does since rf does not effect the linking with L. Suppose we know that $g_k(J_1)$ links L. Use an isotopy on $g_k(A_1)$ to shove $g_k(J_1)$ so close to X_i that $g_{k+1} = g_k$ on the moved curve. Then shove the curve back out to $g_{k+1}(J_1)$ with an isotopy on $g_{k+1}(A_1)$. There is an integer k_0 such that if $k > k_0$, $g_k = g_{k+1}$ on J_1. Hence $g(J_1)$ links L and $rfg(J_1)$ does. □

XII.4. Reducing components. We continue seeking a modification of S^2 to simplify $S^2 \cap \Delta^2$. We would like for $S^2 \cap \Delta^2$ to be so loose that no component of it contains three points on Bd Δ^2. We make use of Theorem XII.2.B and suppose $S^2 \cap \Delta^2$ already has the no-triod property. As was the case in the proof of Theorem XII.2.A, our primary technique in proving Theorem XII.4.A will involve spinning about axes.

THEOREM XII.4.A. *Suppose $S^2 \cap$ Bd Δ^2 is finite, Bd Δ^2 pierces S^2 at each point of $S^2 \cap$ Bd Δ^2, $S^2 \cap \Delta^2$ has the no-triod property, and p_1, p_2, p_3 are three points of $S^2 \cap$ Bd Δ^2 such that each expansion of $S^2 \cap$ Bd Δ^2 in Δ^2 with respect to $S^2 \cap$ Int Δ^2 contains a spanning arc p_1p_2 of Δ^2 and a spanning arc p_2p_3 of Δ^2. Then for each neighborhood N of p_2, there is a homeomorphism $\beta: R^3 \to R^3$ and an expansion E of $\beta(S^2) \cap \Delta^2$ in Δ^2 with respect to $\beta(S^2) \cap$ Int Δ^2 such that*

E does not contain both a spanning arc from p_1 to p_2 and one from p_2 to p_3 in Δ^2,

β is fixed outside N and in a neighborhood of Bd Δ^2,

β is PL, and

$\beta(S^2) \cap$ Bd $\Delta^2 = S^2 \cap$ Bd Δ^2.

PROOF. Figure III.4.A.a illustrates an expansion E' of $S^2 \cap \Delta^2$ that shows that $S^2 \cap \Delta^2$ has the no-triod property. Let $p_1a_1p_2$ be the spanning arc of Δ^2 in E closest to p_3. (No other spanning arc from p_1 to p_2 in E' separates any point of $p_1a_1p_2$ from p_3 in Δ^2.) See 2.J of Appendix. Let p_2xa_1 be an arc in Δ^2 as shown such that $p_2xa_1 \cap (S^2 \cup \text{Bd } \Delta^2) = \{p_2\}$ and p_2xa_1 is PL mod$\{p_2\}$. Let a_2 be a

Schulz

point of Bd Δ^2 near p_2 and in the same component of $R^3 - S^2$ as a_1. The fact that Bd Δ^2 pierces S^2 at p_2 implies that there is an a_2. If there were an arc a_1, a_2 in $\Delta^2 - S^2$ with a_2 to the left of p_2 as shown in Figure XII.4.A.a, then $p_1 a_1 \cup a_1 a_2$ would show that there is an expansion of $S^2 \cap \Delta^2$ which contains no spanning arc from p_1 to p_2. If a_2 is to the right of p_2, then $p_1 a_1 \cup a_1 a_2$ would separate p_2 from p_3 in Δ^2. Hence we seek a β and an $a_1 a_2$ such that $a_1 a_2 \cap \beta(S^2) = \varnothing$. Let $a_1 y a_2$ be a PL arc from a_1 to a_2 in $R^3 - S^2$ such that $a_1 y a_2 \cup a_2 p_2 \cup p_2 x a_1$ is a simple closed curve.

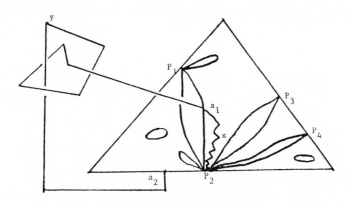

FIGURE XII.4.A.a

Let P^3 be a small polyhedron cube in N such that $P^3 \cap$ Bd Δ^2 is a straight line interval $w p_2 z$ spanning P^3 and missing a_2 such that $p_2 x a_1 \cap$ Bd $P^3 = \{x\}$, $\Delta^2 \cap$ Bd P^3 is arc $wxbz$, and $P^3 \cap a_1 y a_2 = \varnothing$. See Figure XII.4.A.b. Let $z = p_2 a_2 \cap$ Bd P^3.

So that we can speak of the top, bottom, and sides of P^3, we show it as a prism —but this is not required. We suppose S^2 misses the top and bottom of P^3.

We now show that there is a PL arc $xebz$ on Bd $P^3 - S^2$ from x to z. Change the simple closed curve $a_1 y a_2 z p_2 x a_1$ near p_2 to get a simple closed curve missing S, linking the boundary of the top end of P^3 and intersecting Bd P^3 only in $\{z\}$ and $\{x\}$. This closed curve bounds a PL 2-manifold M^2 in $R^3 - S^2$ and M^2 can be used as in the proof of Theorem XII.3.A to get a PL arc $xebx$ in Bd $P^3 - S^2$. We suppose $xebz$ misses the bottom of P^3 and is chosen so that there is a PL isotopy of Bd P^3 onto itself that is fixed on the top and bottom of P^3 and even near x that takes $zbex$ to zbx. Let β be a space PL space homeomorphism invariant on Bd P^3 fixed on $p_2 x a_2$ which takes zbx to $zbex$. Then $\beta(a_1 xbza_2) = a_1 xebza_2$ misses S^2 and if E' is an expansion of $\beta(S) \cap \Delta^2$ that misses $a_1 xbza_2$ in N and agrees with E outside N, E' does not contain both a spanning arc from p_1 to p_2 and one from p_2 to p_3. □

Note that β is the identity except near Bd P^3, and P^3 could have been picked so that each Bd $P^3 \cap rf(B_i^3) = \varnothing$. Hence we suppose that if $S^2 = \alpha rfg(\overline{M_0})$, each $T^2 \cap \beta \alpha rf(B_i^3) = \varnothing$. Hence we are referring to the B_i^3's of §XI.6 rather than the 3-balls G_i used in §XII.1 to describe expansions in $R^3 - Y$.

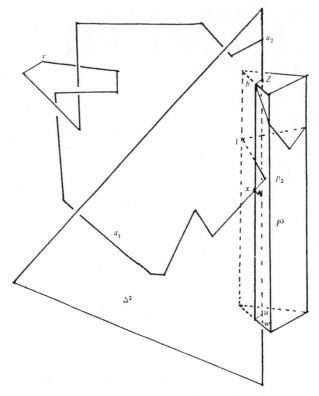

FIGURE XII.4.A.b

THEOREM XII.4.B. *Suppose $S^2 \cap \Delta^2$ is finite, Bd Δ^2 pierces S^2 at each point of $S^2 \cap$ Bd Δ^2, $S^2 \cap \Delta^2$ has the no-triod property, and some component of $S^2 \cap \Delta^2$ contains three points of $S^2 \cap$ Bd Δ^2. Then for each open subset U of R^3 containing $S^2 \cap$ Bd Δ^2 there is a PL homeomorphism $\beta: R^3 \to R^3$ such that*

some triple of points of $S^2 \cap$ Bd Δ^2 that belongs to a component of $S^2 \cap \Delta^2$ fails to belong to any component of $\beta(S^2) \cap \Delta^2$,

if two points belong to different components of $S^2 \cap \Delta^2$ then the two points belong to different components of $\beta(S^2) \cap \Delta^2$, and

β is fixed outside U and in a neighborhood of $S^2 \cap$ Bd Δ^2.

PROOF. Let $A_1, A_2, \ldots, A_n$ be spanning arcs of Δ^2 such that the A_i's miss S^2 and separate Δ^2 between any pair of points of $S^2 \cap$ Bd Δ^2 not belonging to the same component of $S^2 \cap \Delta^2$. We shall pick β to be the identity on $\cup A_i$, and this will insure that two points of $S^2 \cap$ Bd Δ^2 belong to different components of $\beta(S^2) \cap \Delta^2$ if they belong to different components of $S^2 \cap \Delta^2$.

Let E be an expansion of $S^2 \cap \Delta^2$ in Int Δ^2 which contains no-triod and is such that no two points of Bd Δ^2 belong to the same component of E, unless they belong to the same component of $S^2 \cap \Delta^2$. We shall use E to build a triod λ such that λ lies except for its ends in Int $\Delta^2 - S^2$, the ends of λ all belong to $S^2 \cap$ Bd Δ^2, and the ends of λ belong to the same component of $S^2 \cap \Delta^2$.

It follows from the fact that each component of E is locally connected that $E - \text{Bd } \Delta^2$ has only a finite number of components, each with more than one boundary point on Bd Δ^2. Denote these components by $E_1, E_2, \ldots, E_n$. Since E does not have the no-triod property, no $\bar{E}_i$ has more than two points on Bd Δ^2. It follows from repeated use of 2.C of the Appendix that there is a collection of disks $D_1^2, D_2^2, \ldots, D_n^2$ in Δ^2 such that $E_i \subset \text{Int } D_i^2$, $S^2 \cap \text{Bd } D_i^2 = \bar{E}_i \cap \text{Bd } \Delta^2$, and $D_i \cap D_j \cap \text{Int } \Delta^2 = \varnothing$ for $i \neq j$.

We combine certain D_i^2's. If pq is a spanning arc of Δ^2 in E, let $D^2(p, q)$ be the minimal disk in Δ^2 which contains the union of the D_i^2's which contain both p and q. Since some component of E has three points on Bd Δ^2, there are three points p_1, p_2, p_3 of $E \cap \text{Bd } \Delta^2$ so that $D^2(p_1, p_2), D^2(p_2, p_3)$ share the point p_2. In fact, the $D^2(p, p_2)$'s fan out about q_2 so that for some pair of $D^2(p, q)$'s (say $D^2(p_1 p_2), D^2(p_2, p_3)$), there is an arc A in Int Δ^2 from $D^2(p_1, p_2)$ to $D^2(p_2, p_3)$ that does not intersect any other $D^2(p, q)$. The arc A might intersect S^2, but by rerouting it around small components of $E \cap \text{Int } \Delta^2$ we obtain an arc A' in Int $S^2 - S^2$ from $D^2(p_1, p_2)$ to $D^2(p_2, p_3)$. Then λ is a triod in Bd $D^2(p_1, p_2)$ $\cup$ Bd $D^2(p_2, p_3) \cup A'$.

The proof is finished by a modification of the techniques of the proof of Theorem XII.4.A. One simultaneously gets a PL homeomorphism $\beta: R^3 \to R^3$ and a triod λ' in $\Delta^2 - \beta(S^2)$ such that β is fixed in a neighborhood Bd Δ^2 and except near the ends of λ while the ends of λ' are on Bd Δ^2 and near the ends of λ. The triod λ' prevents any component of $\Delta^2 \cap \beta(S^2)$ from containing the three points. $\square$

We point out a weakness of Theorems XII.4.A and XII.4.B. They did not claim that $\beta(S^2) \cap \Delta^2$ had a no-triod property. If we knew that it always did, a finite number of applications of Theorem XII.4.B would yield an S^2 such that no three points of $S^2 \cap \text{Bd } \Delta^2$ belong to some component of $S \cap \Delta^2$. However, we can prove the following result with an oscillating argument which uses a center push at alternate stages.

Repeated applications of Theorem XII.4.B, with intermediate applications of Theorem XII.2.A to recover the no-triod property, gives the following result.

THEOREM XII.4.C. *Suppose $S^2 \cap \Delta^2$ is finite. Then there is a PL homeomorphism $\gamma: R^3 \to R^3$ such that γ is the composition of a finite number of α's as described in Theorem XII.2.A and β's as described in Theorem XII.4.B such that*

no component of $\Delta^2 \cap \gamma(S^2)$ has three points of Bd Δ^2,
γ is the identity on T^1,
γ is near the identity on R^3, and
if $S^2 = rfg(\overline{M_0})$ of §§XI.4 and XI.5, then the $\gamma rf(B_i^3)$'s miss T^2. $\square$

If Theorem XII.4.C is applied to each 2-simplex of T intersecting S^2 one obtains the following result.

THEOREM XII.4.D. *Suppose $S^2 \cap T^1$ is finite. Then there is a PL homeomorphism $\gamma: R^3 \to R^3$ such that*

if Δ^2 is a 2-simplex of T, no component of $\Delta^2 \cap (S^3)$ has three points on Bd Δ^2,

γ is the identity on T^1,

γ is near the identity on R^3, and

the $\gamma rf(B_i^3)$'s miss T^2.   □

XII.5. Intersection of T^2 with modification of K^2. Suppose we have used h to change K^2 to S^2 and γ to change S^2 to Z^2. Then $Z^2 = \gamma(S^2) = \gamma h(K^2) = \gamma rfg(\overline{M_0}$. If this is done properly, we have the following.

$T^1 \cap S^2 = T^1 \cap Z^2$,

γrfg is near the identity on $\overline{M_0}$,

the $\gamma rf(B_i^3)$ miss T^2.

Why are we obcessed with the location of the $\gamma rf(B_i^3)$'s? The reason is that if someone cuts some small mutually disjoint holes in a 2-sphere and someone else does the same, then what is left might not be a 2-sphere with small mutually disjoint holes but rather a 2-sphere with big holes. In Chapter XIII we build a hollow ball $\phi(K^2 \times [-1, 1])$ that contains most of $\overline{M_0}$ and $\overline{M_0}$ contains most of K^2. How do we show $\phi(K^2 \times [-1, 1])$ contains most of K^2? We use a suitable selection of the $\gamma rf(B_i^3)$'s, but there may be a better way.

THEOREM XII.5. *As in Theorem XI, suppose that K^2, T, the C_i^3's, g, M_0, r, f are such that each component of $K^2 - \overline{M_0}$ has diameter less than ε, $d(rfg, \mathrm{Id}) < \varepsilon$ on $\overline{M_0}$, and each diameter $C_i^3 < \varepsilon$. Also, γ is a PL homeomorphism of R^3 into R^3 as suggested in Theorem XII.4.D such that $d(\gamma rfg, \mathrm{Id}) < \varepsilon$ on $\overline{M_0}$ and γ is the identity on the $rf(B_i^3)$'s. If Q is a closed subset of $T^2 \cap \gamma rfg(\overline{M_0})$ such that each component of $\gamma rfg(\overline{M_0}) - Q$ has diameter less than ε and U is an open subset of R^3 containing Q, then there is a PL ε-homeomorphism $H: R^3 \to R^3$ such that each component of $H(K^2) - U$ has diameter less than 7ε.*

PROOF. The proof resembles that of Theorem XI.6. For some large i, $\gamma r_i f_i$ is a suitable H.   □

CHAPTER XIII

SIDE APPROXIMATION THEOREM

A theorem which served as a breakthrough to enable us to use PL techniques to determine properties of topological surfaces in 3-manifolds is the following.

THEOREM XIII.A (APPROXIMATION THEOREM). *Suppose K^2 is a Topological 2-sphere in a PL 3-manifold M^3. Then for each $\varepsilon > 0$ there is a homeomorphism h of K^2 onto a PL 2-sphere $h(K^2)$ in M^3 such that $d(h, \mathrm{Id}) < \varepsilon$.* $\square$

Another version of the approximation theorem is the following.

THEOREM XIII.B (SIDE APPROXIMATION THEOREM). *Each 2-sphere K^2 in R^3 can be approximated almost from either side by a PL 2-sphere—that is, for each $\varepsilon > 0$ and each component U of $R^3 - K^2$ there is a homeomorphism h of K^2 onto a PL 2-sphere such that*

(1) $d(h, \mathrm{Id}) < \varepsilon$ and

(2) $h(K^2)$ contains a finite collection of mutually disjoint ε-disks $D_1, D_2, \ldots, D_n$ such that $h(K^2) - \cup D_i \subset U$. $\square$

A stronger version of Theorem XIII.B is the following.

THEOREM XIII.C. *Suppose K^2 is a topological 2-sphere in R^3. Then there is a homeomorphism $\phi : K^2 \times [-1, 1] \Rightarrow R^3$ such that*

$\phi(x \times [-1, 1])$ lies in a small neighborhood of x,

each component of $K^2 - \phi(K^2 \times [-1, 1])$ is small, and

each of $\phi(K^2 \times \{-1\})$ and $\phi(K^2 \times \{1\})$ is PL. $\square$

Bing gave the first proof of the approximation theorem in [B₉], and both Bing [B₁₆, B₂₂] and Cannon [C₅] have given proofs of the side approximation theorem. Lister gives a version in [L₂]. There are variations of these theorems as they apply to surfaces rather then spheres, to positive ε functions rather than to positive ε constants, and to PL 3-manifolds rather than R^3. Proofs of these variations are frequently similar to the proof which we shall give of Theorem III.C. There have been many applications of these theorems, and we give some in Chapter XVIII. In this chapter we restrict ourselves to the consideration of the proof of Theorem XIII.C since the other two follow from similar methods.

The original proofs of Theorem XIII.C were exercises in epsilonics, and δ's were picked so that we came out at the end with the proper prescribed ε's. We shall not do that here. Rather, we shall take certain sets as "small" and certain maps as "close" and leave it to the reader to see that one can define these so as to get the proper conclusions. Experience suggests that many people are more interested in the ideas of a proof than in the epsilonics.

In this vein we say that a set *contains most of a* 2-sphere if it contains all of the 2-sphere except for a subset that lies in the union of a finite collection of small mutually disjoint disks. We sometimes say that the set contains a 2-sphere with small holes. It is understood that we have already decided on the meaning of "small" and then picked the near-covering set after this determination is made.

In the same vein we can speak of *almost approximating* a 2-sphere K^2 from the components of its complement. The Alexander horned sphere described in §IV.3 can be approximated from its interior but only almost from its exterior. The same holds for Antoine's wild sphere described in §IV.8 and the Fox-Artin spheres described in §IV.9. However, the wild 2-sphere which is the common boundary of two pieces of S^3 as mentioned in Theorem IV.6.A cannot be approximated from either side. It can be almost approximated from either side.

In building $\phi : K^2 \times [-1, 1]$ we make extensive use of a triangulation T of small mesh. One of the reasons we want T to have small mesh is so that if ϕ' is a homeomorphism of K^2 into R^3 that is near the identity and X is a subset of $\phi'(K^2)$ that is near a 3-simplex of T, we want $\phi(K^2) - X$ to have only one big component. The nonhour glass property 3.C from the Appendix gives such a triangulation T. We use T^i to denote the i-skeleton of T.

If $\{X_\alpha\}$ is an indexed collection of sets we use $\cup X_\alpha$ to denote their union. If the elements are not indexed and X is a typical one, we denote their union by $\cup X$.

XIII.1. Very special case. To give a flavor to the type of proof of Theorem XIII.C to be given we prove the theorem under a simple set of conditions. Although we shall not reduce the general case to this situation, it does provide a hint as to how to proceed. We first prove the theorem in the very special case that

$K^2 \cap T^0 = \varnothing$,

$K^2 \cap T^1$ is finite,

T^1 pierces K^2 at each point of $K^2 \cap T^1$, and

each component of $K^2 \cap \Delta^2$ is a spanning arc of Δ^2 for each 2-simplex Δ^2 of T.

The proof in this very special case is given in four steps. A local view of $K^2 \cap T^2$ is given by Figure XIII.1.A.

1. In this step we consolidate a subdivision of K^2 provided by $K^2 \cap T^2$. Denote the components of $K^2 - T^2$ by D's. Each D is small since it lies in a 3-simplex of T. It would be convenient if each D were an open disk, but feelers of

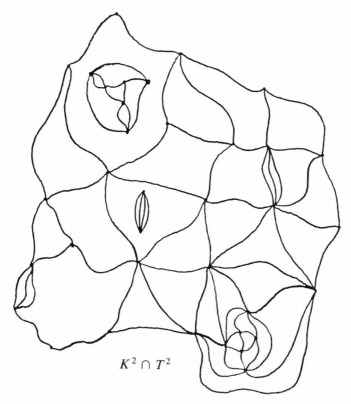

$K^2 \cap T^2$

FIGURE XIII.1.A

K^2 may cause some D's to be open disks with holes. In this case we fill in the small holes and let E be the closure of resulting disk. The complement of E is a large open disk on K^2 while Bd $E \subset$ Bd D and lies on the boundary of some 3-simplex of T. Note that E is small since Bd E is. Let $E_1, E_2, \ldots, E_m$ be the E's that are not a subset of other E's. Then $E_1, E_2, \ldots, E_m$ provides a partitioning of K^2 such that if two of the E_i's intersect each other, the intersection lies on the boundary of each. Then $G = \cup$ Bd E_i is a connected finite graph (not necessarily PL) without separating points and each component of $K^2 - G$ is an Int E_i.

2. In this step we start defining ϕ by describing it on $p \times [-1, 1]$ for points $\{p\}$ of $G \cap T^1$. Let $B_1, B_1, \ldots, B_r$ be straight arcs on $T^1 - T^0$ such that the B's are mutually disjoint and each point $p \in G \cap T^1$ is the center of one of the B's. Then ϕ takes $p \times [-1, 1]$ linearly onto the B centered at p so that $\phi(p \times (-1)) \in$ Int K^2 and $\phi(p \times 1) \in$ Ext K^2.

3. In this step we extend ϕ to $G \times [-1, 1]$. If A_i is a component of $G \cap \Delta^2$, we consider a PL disk C_i in the Δ^2 containing A_i as shown in Figure XIII.1.B. The disk C_i contains A_i; $C_i \cap$ Bd Δ^2 is the union of two B's; pairs of C's on the same Δ^2 are mutually disjoint. The map ϕ is extended to a homeomorphism taking each $A_i \times [-1, 1]$ onto the PL disk C_i.

4. On this final step of the very special case we extend ϕ to $E_i \times [-1, 1]$. Each component of the union of the C_i's on the boundary of any 3-simplex of T is a PL

Figure XIII.1.B.

annulus, and ϕ has already been defined to take $G \times [-1, 1]$ into the union of such annuli. For each 3-simplex Δ^3 of T consider all such annuli on Bd Δ^3 and pick a family of mutually disjoint PL 3-cells in Δ^3 such that each annulus is the intersection of one of the 3-cells and Bd Δ^2. Then ϕ is extended to take $E \times [-1, 1]$ onto the appropriate 3-cells. Chapter XIV explains how to get such an extension.

The reason that K^2 lies almost in $\phi(K^2 \times [-1, 1])$ is that $G \subset \cup C_i \subset \phi(K^2 \times [-1, 1])$. □

XIII.2. Modification of very special case. In the general case we shall deal with a situation where a component of $K^2 \cap \Delta^2$ contains many spanning continua whose intersections with Int Δ^2 are mutually disjoint. We decided to confront the difficulties of such a situation by examining it in a simple environment—namely, we alter the restrictions in the very special case by replacing the requirement that each component of $K^2 \cap \Delta^2$ is a spanning arc by the less restrictive condition that each component of $K^2 \cap \Delta^2$ intersects Bd Δ^2 in just two points and is the union of a finite number of spanning arcs of Δ^2 whose interiors are mutually disjoint.

In the first step of the proof of the very special case we consolidated D's to get E's and let $G = \cup$ Bd E. To get a substitute for the D's we let Δ^2 range over the 2-simplexes of T and denote the set of components of $(K^2 \cap \Delta^2)$'s by W's. Then $\cup W = K^2 \cap T^2$ and $\cup W$ subdivides K^2 into a finite number of small pieces which we call D's.

We now seek to consolidate the D's by reducing each W to a spanning arc. The problem lies in selecting the spanning arcs since if A_{1i} is a spanning arc of Δ_1^2 in K^2 there may be other A_{1j}'s with the same ends and in the same Δ_1^2. Even worse, there may be other A_{kj}'s in K^2 with both ends in common, yet some of them lie in different Δ_i^2's. See Figure XIII.2. We need to pick G so that no more than one A_{kj} is picked in each W and each Bd E_i lies in a 3-simplex of T.

We let E_1 be a small disk in K^2 bounded by a certain subset of some $K^2 \cap$ Bd Δ^3. Also E_1 is maximal in that it is the closure of the union of certain

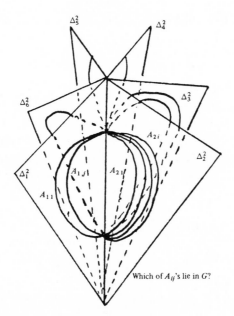

FIGURE XIII.2

pieces of $K^2 - T^2$, Bd E_1 lies on some Bd Δ_i^3, diameter E_1 is small, but E_1 is not a subset of any other such disk. It may be shown that if Bd E_1 contains an arc in a W, E_1 contains all of the W.

We let E_2 be a maximal small disk in $K^2 - \text{Int } E_1$ bounded by a subset of some $(K^2 - \text{Int } E_1) \cap \text{Bd } \Delta^3$. If Bd E_2 contains an arc in a W, $W \subset E_1 \cup E_2$.

In a similar fashion we let E_3 be a maximal small disk in $K^2 - (\text{Int } E_1 \cup \text{Int } E_2)$ whose boundary is a subset of some $K^2 \cap \text{Bd } \Delta^3$. In a finite number of such steps we have a subdivision of K^2 into E_i's such that each Bd E_i lies on a $T^2 \cap \text{Bd } \Delta^3$ but none lies on a Δ^2. Hence we associate with each E_i a 3-simplex of T—namely, the one containing its boundary.

We let $G = \cup \text{Bd } E_i$ and proceed as in the case of the very special case to define $\phi(K^2 \times [-1, 1])$. Note that G contains at most one spanning arc from each W. The situation shown in Figure XIII.2 might prevent some W's from contributing any arcs to G.

XIII.3. Special case. We repeat that the very special case and its modification were discussed primarily from heuristic reasons. The present special case is used in the proof of the side approximation theorem. The fact that the theorem holds in this special case is used in the proof of the general case. In the special case we suppose

$K^2 \cap T^0 = \varnothing$,

$K^2 \cap T^1$ is finite,

T^2 pierces K^2 at each point of $K^2 \cap T^1$,

each component of $K^2 \cap \Delta^2$ that intersects Bd Δ^2 intersects it in precisely two points, and

each $K^2 \cap \Delta^3$ lies on the interior of a small disk in K^2.

We shall show in this special case that there is a homeomorphism $\phi: K^2 \times [-1, 1] \to R^3$ such that

$\phi(x \times [-1, 1])$ lies in a small neighborhood of x,

Bd $\phi(K^2 \times [-1, 1])$ is PL, and

there is a component X of $K^2 \cap T^2$ such that $X \subset \phi(K^2 \times [-1, 1])$ and each component of $K^2 - X$ lies in small disk in K^2.

The components of $K^2 - X$ can be obtained by filling in the small holes of the components of $K^2 - T^2$. The following paragraph explains the procedure.

If F is a small disk in K^2 whose boundary misses T^2, we add F to each component of $K^2 - T^2$ containing Bd F. We follow this procedure of enlarging components by considering all such F's. An enlarged component need no longer lie in a 3-simplex of T but its boundary does. As the components enlarge, the complement in K^2 of their union contracts. The set X is the remainder of $K^2 \cap T^2$ after the F's are removed. It may be shown that X is the big component of $K^2 \cap T^2$.

In the very special case the components of $K^2 \cap \Delta^2$ were arcs, but here in the special case the components of $K^2 \cap \Delta^2$ are continua. One of our tasks is to pick arcs near the continua. However, before selecting them we build $\phi(K^2 \times [-1, 1])$. We build it as a PL union of PL 3-balls rather than as a map. For each point $p \in X \cap T^1$ we define $\phi(p \times [-1, 1])$ as done in Step 2 of the very special case. If X_i is a component of $X \cap \Delta^2$, we shall associate an arc A_i in K^2 near X_i and follow Step 3 of the very special case to send a PL disk $C_i = \phi(A_i \times [-1, 1])$ into Δ^2. Instead of C_i containing A_i as shown in Figure XIII.1, it contains X_i. We delay until later the selection of the A_i. However, we do sew in PL 3-cells along annuli as suggested by Step 4 of the very special case to complete the PL set $\phi(K^2 \times [-1, 1])$. Note that $\phi(K^2 \times [-1, 1])$ contains X so each component of $K^2 - \phi(K^2 \times [-1, 1])$ lies in a small disk in K^2.

We return to the task of picking the substitutes for the W's, A's, and G's. To that end we expand each $K^2 \cap \Delta^2$ in $K^2 -$ Bd Δ^2 so that if W^+ is the resulting expansion (as described in §XII.1), each point of $W^+ -$ Bd Δ^2 is nearer to Δ^2 than to $T^2 -$ Int Δ^2, each component of W^+ that intersects Bd Δ^2 intersects it in just two points p, q, and W^+ lies in the preselected $\phi(K^2 \times [-1, 1])$. Some components of W^+ may fail to intersect X and we ignore them. Let X_i^+ be the component of W^+ containing X_i. Use X_{ij}^+ to denote the closure of a component of $X_i^+ - (\{p\} \cup \{q\})$ and X_{ij}^{++} to denote the disk in K^2 formed by filling small holes of X_{ij}^+. Ignore X_{ij}^{++}'s that lie in others and use Y_{ij}^{++} to denote X_{ij}^{++}'s with both p and q on its boundary. Select a spanning arc A_{ij} of X_{ij}^{++} from p to q in X_i^+ and let the union of these A_{ij}'s in X_i^+ be a W. Since we do not have $W^+ \subset \Delta^2$ we do not claim $W \subset \Delta^2$. However, W^+ "is near" Δ^2, W "is near" Δ^2, and A_{ij} "is near" Δ^2. In retrospect we see that we obtained arcs near X by expanding X and taking arcs in the expansion.

Then $\cup W$ chops up K^2 in somewhat the same way that $\cup W$ did in the modification of the very special case—the main difference being that the boundary of an element D of the subdivision does not necessarily lie on a Bd Δ^3. We now show that the component of $K^2 - \cup W$ are small and the boundary of each component of $K^2 - \cup W$ is near some Bd Δ^3.

Consider the components of each of $K^2 - T^2$, $K^2 - X$, $K^2 - \cup X_{ij}^{++}$, $K^2 - \cup Y_{ij}^{++}$. We have already noted that components of $K^2 - T^2$ lie in small disks since they lie in 3-simplexes of T and components of $K^2 - X$ lie in small disks since they are obtained by filling in small holes in components of $K^2 - T^2$. The components of $K^2 - \cup X_i^+$ lie in small disks since $\cup_j X_i^+$ is an expansion of X in $K^2 - T^1$.

We continue changing sets in K^2. Each components of $K^2 - \cup X_{ij}^{++}$ lies in a small disk. Since $\cup X_{ij}^{++}$ is the union of $\cup Y_{ij}^{++}$ and X_{ij}^{++}'s each of which has one point of T^1 on its boundary, components of $K^2 - \cup Y_{ij}^{++}$ lie in small disks. Also, if U is a component of $K^2 - \cup Y_{ij}^{++}$, it may be shown that Bd U is a simple closed curve which lies except for a finite set of points in a component of $K^2 - T^2$.

Finally we replace each Y_{ij}^{++} with an arc A_{ij} in W^+ that spans Y_{ij}^{++} see Figure XIII.3. Since each of the small disks shown in Figure XIII.3 is small, components of $K^2 - W$ lie in disks whose diameters are not more than three times as big. In fact, since the dotted curve is near the boundary of a Δ^3, it lies in a small disk, and we do not need a multiple of 3. Instead of a component of $K^2 - \cup W$ having its boundary lie in a Δ^3, the boundary of the component "is near" the boundary of a Δ^3.

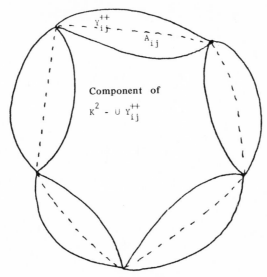

FIGURE XIII.3

Following the methods used to get E_1, E_2,...,E_n in the modification of the very special case we let E_1 be a small maximal disk in K^2 such that Bd $E_1 \subset \cup W$ and for some $\Delta^3 \in T$, Bd E_1 "is near" Bd Δ^3. Continuing we get a subdivision E_1, E_2,...,E_n where instead of supposing the Bd E_i is in Bd Δ^3, we suppose it "is near" Bd Δ^3. Also $\cup$ Bd $E_i = G = \cup A_j$. Each component of $K^2 - G$ lies in a small disk in K^2.

Let $A_i = pq$ be an arc in G which is near a Δ^2. It lies in K^2 but need not lie in this Δ^2. The PL disk $C_j = \phi(A_i \times [-1, 1])$ does lie in Δ^2 and contains the component of $K^2 \cap \Delta^2$ containing $\{p\} \cup \{q\}$. Then $\phi(A_i)$ is any spanning arc of C_i from p to q. Also $\phi(A_i \times [-1, 1]) = C_i$ and $\cup \phi(\overline{E_i} \times [-1, 1])$ is a PL 3-cell in a Δ^3 that intersects Bd Δ^3 in an annulus.

It may be noted that we have indulged in a bit of overkill. We wanted $\phi(K^2 \times [-1, 1])$ to almost contain K^2. We did this by causing $\phi(K^2 \times [-1, 1])$ to contain the big component X of $K^2 \cap T^2$. It would have been enough to insist only that $\phi(K^2 \times [-1, 1])$ contain enough of $K^2 \cap T^2$ to chop K^2 into small pieces. $\square$

XIII.4. General case. In this final section of Chapter XIII we make use of the K^2, ε, C_i^3's, g, M_0, f, r, h, B_i^3's, Z^2 used in Chapters XI and XII.

One might attempt to handle the proof of the general case of Theorem XIII.C by using Theorem XI.2, XII.3.B, and XII.4.D to get a $\gamma h : k^2 \to Z^2$ such that γh is near the identity and Z^2 satisfies the following conditions:

$Z^2 \cap T^2 = \varnothing$,

$Z^2 \cap T^1$ is finite,

T^1 pierces Z^2 at each point of $T^1 \cap Z^2$,

each component of $Z^2 \cap \Delta^2$ that intersects Bd Δ^2 intersects it in precisely two points.

The special case gives a $\phi(Z^2 \times [-1, 1])$. The gap in the proof for the general case is that our argument shows that Z^2 (rather than K^2) almost lies in $\phi(Z^2 \times [-1, 1])$. We could dodge the gap if γh were a PL homeomorphism of R^3 onto R^3 near the identity since K^2 would almost lie in $(\gamma h)^{-1}\phi(Z^2 \times [-1, 1])$. However, h is not defined on R^3. Nevertheless we can get around the difficulty by using Theorem XII.5.

We use certain parts of the crippled proof given above. We get γh and Z^2 as suggested there where $h(K^2) = S^2 = rfg(\overline{M}_0)$ and each $T^2 \cap \gamma rf(B_i^3) = \varnothing$. The mesh of T is small so that if Q is the big component of $Z^2 \cap T^2$ components of $Z^2 - Q$ are small. We pick an $\varepsilon > 0$ and resurrect the following properties from Chapters XI and XII.

> Each component of $K^2 - M_0$ lies in an ε-disk in K^2, each $K^2 \cap \Delta^3$ lies in an ε-disk in K^2, the $K^2 \cap C_i^3$'s lie in ε-disks of K^2, each component of $Z^2 - Q$ lies in an ε-disk of Z^2, and $d(\gamma h, \text{Id}) < \varepsilon$.

So as not to confuse the ϕ used in §XIII.3 with the one we seek to insure the truth of Theorem XIII.C, we designate the one given in §XIII.3 by ϕ'. Then §XIII.3 guarantees a homeomorphism $\phi' : (Z^2 \times [-1, 1]) \Rightarrow R^3$ such that

$Q \subset \phi'(Z \times (-1, 1))$,

$d(z, \phi'(z \times t)) < \varepsilon$ for $t \in [-1, 1]$, and

Bd $\phi'(Z^2 \times [-1, 1])$ is PL.

Let $U = \phi'(Z^2 \times (-1, 1))$ and H be the PL ε-homeomorphism of R^3 onto itself promised by Theorem XII.5 such that each component of $H(K^2) - U$ has diameter less than 7ε. A suitable homeomorphism $\phi : K^2 \times [-1, 1] \Rightarrow R^3$ is given by

$$\phi(x \times t) = H^{-1}\phi'(\gamma h(x) \times t).$$

We note that if $x \in K^2$, $d(x, \gamma h(x)) < \varepsilon$, $d(\gamma h(x), \phi'(\gamma h(x) \times t)) < \varepsilon$, and $d(H^{-1}, \mathrm{Id}) < \varepsilon$. Hence $d(x, \phi(x \times t)) < 3\varepsilon$ and $\phi(x \times [-1, 1])$ lies in a small neighborhood of x.

Theorem XII.5 gives that each component of $H(K^2) - U$ is of diameter less than 7ε. Apply H^{-1} and find that each component of $H^{-1}(H(K^2) - U)$ is of diameter less than 9ε. But $H^{-1}(H(K^2)) = K^2$ and

$$H^{-1}(U) = H^{-1}(\phi'(Z^2 \times (-1, 1))) = H^{-1}\phi'(\gamma h(K^2) \times (-1, 1))$$
$$= \phi(K^2 \times (-1, 1))$$

and a component of $K^2 - (\phi(K^2) \times (-1, 1))$ is a component of

$$H^{-1}(H(K^2) - U)$$

and is hence small.

Since H is PL and each Bd $\phi'(Z^2 \times [-1, 1])$ is PL, each of $\phi(K^2 \times -1)$ and $\phi(K^2 \times 1)$ is PL. See Figure XIII.4.

We used two concepts to indicate that a subset of a 2-sphere K^2 is small. Sometimes we said that the subset had a small diameter (was an ε-set), and at other times we used the more restrictive requirement that the subset lies in a small

$\phi'(Z^2 \times [-1, 1])$ $H(K^2)$ Z^2

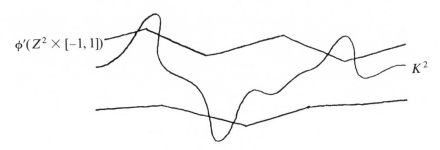

$\phi'(Z^2 \times [-1, 1])$ K^2

FIGURE XIII.4

disk (ε-disk) in K^2. Since we used Theorem XII.5 which dealt with ε-sets, we concluded that components of $K^2 - \phi(K^2 \times (-1, 1))$ were 9ε-sets. However, we actually had a sequence of these ϕ's with the associated 9ε's approaching 0. Property 3.B of the Appendix implies that subsets of K^2 lie in small disks if they are very small subsets. Hence we can pick a ϕ (perhaps different from the first considered) so that components of $K^2 - \phi(K^2 \times (-1, 1))$ lie in small disks. $\square$

Are we creatures of habit? In the very special case we used arcs A_i which were components of the $K^2 \cap \Delta^2$'s and subdivided K^2 by using the union of the A_i's. From our acquaintance with triangulations we have learned how to use the 1-skeleton to divide a 2-sphere into small pieces and by using the Schoenflies theorem have learned how to extend a homeomorphism on the 1-skeleton to a homeomorphism on the 2-sphere. But is there a better way of subdividing a 2-sphere? In a modification of the very special case we ignored some of the A_i's and just kept one from each batch. In this case it was not that we could not handle all of them but it would have meant more trouble for small gain. In the special case we considered components of $K^2 \cap \Delta^2$ that intersected Bd Δ^2 in 2 points and used a nearby W in K^2 as a substitute for these components. In the general case, we used γh to change K^2 so that the components of $\gamma h(K^2) \cap \Delta^2$ that intersected Bd Δ^2 in two points. Were these moves necessary? It seems natural to expect that if we could understand the situation better, components of $K^2 \cap \Delta^2$ that intersected Bd Δ^2 in several points might nail down a homeomorphism more firmly than just one intersecting Bd Δ^2 in two points. From real life one might observe that the more braces a scaffolding has, the more rigid it becomes.

CHAPTER XIV

THE PL SCHOENFLIES THEOREM FOR R^3

XIV.1. Extending the PL Schoenflies theorem. It was shown in Chapter III that if P is a polygonal simple closed curve in R^2, then there is a PL homeomorphism h of R^2 onto itself such that $h(P)$ is a triangular simple closed curve. In this chapter we prove an analogous result for R^3.

THEOREM XIV.1 (PL SCHOENFLIES THEOREM FOR R^3). *If S is a PL 2-sphere in R^3, there is a PL homeomorphism $h: R^3 \to R^3$ such that $h(S)$ is the boundary of a 3-simplex. Furthermore, if U is an open set in R^3 containing $S \cup$ Int S there is such an h that is the identity outside U.*

Moise gave a proof of the PL Schoenflies theorem for R^3 in [$\mathbf{M}_7$]. His proof was modeled after methods used by J. W. Alexander [$\mathbf{A}_2$] where Alexander considered a weaker version of this theorem. Instead of seeking a PL homeomorphism of R^3 onto itself, Alexander wanted a non-PL homeomorphism that would send S onto the boundary of a round ball. His proof was unclear since he did not discuss the role of the "logarithmic potential". In [$\mathbf{G}_3$], Graub gave a proof of the PL Schoenflies theorem for R^3 using regular neighborhood theory.

Many of the preliminary results of this chapter generalize to higher dimensions, but it is not known that the PL Schoenflies theorem does. There are several possible extensions but one version might be as follows. Suppose Δ^n is an n-simplex and h is a PL homeomorphism of Bd Δ^n into R^n. Can h be extended to a PL homeomorphism of Δ^n into R^n? For $n = 4$, the problem is so elusive that neither a proof nor a counterexample is at hand. William T. Eaton has suggested results in lectures. A solution of such a PL Schoenflies theorem in higher dimensions is one of the roadblocks to learning more about the PL structure of higher-dimensional manifolds.

XIV.2. Pushing disks about on tetrahedra. In proving the PL Schoenflies theorem for R^3 we need to learn how movements on the boundary of a tetrahedron can be extended to R^3.

An open subset of R^3 is a *polyhedral* if it is the union of a finite number of interiors of tetrahedra.

THEOREM XIV.2.A. *Suppose that in R^3*
Δ^3 *is a tetrahedron,*
D *is a polyhedral disk on* Bd Δ^3,
$v_0 v_1 v_2$ *is a triangular disk in D such that $v_0 v_1 v_2 \cap$ Bd $D = v_1 v_2$, and*
U *is a polyhedral open set in R^3 containing $v_0 v_1 v_2 -$ Bd$(v_1 v_2)$.*
Then there is a PL homeomorphism h of R^3 onto itself such that
$\quad h(D) = (D - v_0 v_1 v_2) \cup v_0 v_1 \cup v_0 v_2,$
$\quad h(\Delta^3) = \Delta^3,$
$\quad h$ *is fixed on $R^3 - U$.*

PROOF. There are several cases to be considered in the proof of Theorem XIV.2.A according as to how Bd $v_0 v_1 v_2$ intersects the edges and vertices of Δ^3. Figure XIV.2.A shows the special case where $v_1 v_2$ lies on an edge of Δ^3 and v_0 lies on the interior of a face. We let x be the midpoint of $v_1 v_2$; a be a point of Bd Δ^3 such that v_0 is between x and a, while $a v_1 v_2$ lies except for possibly v_1, v_2 in $D \cap U$; b is a point of Bd Δ^3 such that $b v_1 v_2$ lies except for possibly v_1, v_2 in U and intersects D only in $v_1 v_2$; and x' is the midpoint of bx. Let $p_1 x p_2$ be a short segment piercing Bd Δ^3 at x such that the four tetrahedra with vertices at p_1, p_2 and bases $a v_1 v_2$, $b v_1 v_2$ all lie except for possibly v_1, v_2 in U. The required homeomorphism is fixed outside these four tetrahedra, and for $p = p_1$, p_2 and $i = 1, 2$ sends the tetrahedra $p v_i a x$, $p v_i \times x'$, $p v_i x' b$ linearly onto $p v_i a v_0$, $p v_i v_0 \times p v_i x b$. An examination of h reveals that it is the composition, $h_2 h_1$, of two pushes—the first of which pushes x to v_0 and the second of which pushes $h_1(x')$ to x. The first removes $v_0 v_1 v_2$ and the second takes $h_1(\text{Bd } \Delta^3)$ back to Bd Δ^3. Recall that we defined a *push* in §II.8 for R^2 and here we are using the 3-dimensional generalization.

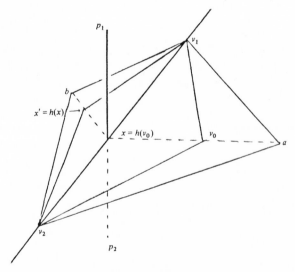

Figure XIV.2.A

If v_0 lies in the interior of a face of Bd Δ^3 and v_1v_2 does not lie on an edge, we can get an h as above which is one push.

If v_0 lies on an edge we start by getting a PL homeomorphism g of R^3 itself such that $g(v_0v_1v_2)$ is a triangular disk, $g(v_0)$ does not lie on an edge, and g is invariant on Bd Δ^3 ($G(\text{Bd } \Delta^3) = \text{Bd } \Delta^3$). Then we use an h as described above to remove $g(v_0v_1v_2)$ from $g(D)$ and use $g^{-1}hg$ as a suitable homeomorphism removing $v_0v_1v_2$ from D. $\square$

The following modification of Theorem XIV.2.A has a similar proof.

THEOREM XIV.2.B. *Theorem* XIV.2.A *follows if the hypothesis that* $v_0v_1v_2 \cap$ Bd $D = v_1v_2$ *is replaced by hypothesis that* $v_0v_1v_2 \cap$ Bd $D = v_1v_0 \cup v_0v_2$. $\square$

Theorems XIV.2.A and XIV.2.B may be extended as follows.

THEOREM XIV.2.C. *Suppose* Δ^3 *is a 3-simplex in* R^3, U *is an open set in* R^3 *containing* Bd Δ^3, *and* D_1, D_2 *are polyhedral disks on* Bd Δ^3. *Then there is a PL homeomorphism* h *of* R^3 *onto itself such that*

$h(D_1) = D_2$,
$h(\Delta^3) = \Delta^3$, *and*
h *is the identity outside* U.

PROOF. The proof is by a shelling argument given by Theorem VI.6.A. Let D_3 be a polyhedral disk in Bd Δ^3 and T_1, T_2, T_3 be triangulations of D_1, D_2, D_3 such that T_1, T_3 share a 2-simplex Δ_1^2 and T_3, T_2 share a 2-simplex Δ_2^2. Theorem VI.6.A shows that D_i can be shelled to any 2-simplex in T_i. It follows from Theorems XIV.2.A and XIV.2.B that there are PL homeomorphisms h_{11}, h_{31}, h_{32}, h_{22} of R^3 onto itself such that for appropriate i, j's, $h_{ij}(D_i) = \Delta_j^2$, $h_{ij}(\Delta^3) = \Delta^3$, and h_{ij} is the identity outside U. The required homeomorphism h is $h_{22}^{-1}h_{32}h_{31}^{-1}h_{11}$. $\square$

XIV.3. Extending a θ curve result. Theorem III.1.B says that if two simple closed curves of a polygonal θ curve in R^2 satisfied a certain Schoenflies property, then the third did also. This section considers a similar result in the next dimension.

We say that a polyhedral 2-sphere S in R^3 is *tetrahedral* if for each open set U in R^3 containing $S \cup$ Int S, there is a PL homeomorphism h of R^3 onto itself such that $h(S)$ is the boundary of a tetrahedron and h is the identity outside U. This definition is of only temporary use since we show in Theorem XIV.1 that any polyhedral 2-sphere in R^3 is tetrahedral.

Suppose that X, Y are two disjoint compact sets in a Euclidean space such that if one segment (straight closed interval) from X to Y intersects another such segment, then the intersection is an end point of each. We recall that the *join* of X and Y is the union of all segments from X and Y.

THEOREM XIV.3.A. *Suppose* S_1, S_2 *are two polyhedral 2-spheres in* R^3 *such that* S_1 *is tetrahedral,* $S_2 \not\subset S_1 \cup$ Int S_1, *and* $S_1 \cap S_2$ *is a disk* D. *Then for each polyhedral open set* U *in* R^3 *containing* $(S_1 \cup$ Int $S_1) -$ Bd D *there is a PL homeomorphism* h *of* R^3 *onto itself such that* $h(S_1 \cup S_2 -$ Int $D) = S_2$ *and* h *is fixed outside* U.

PROOF. Let $v_0v_1v_2v_3$ be a tetrahedron and h_1 be a PL homeomorphism of R^3 onto itself such that $h_1(S_1) = $ Bd $v_0v_1v_2v_3$. It follows from Theorem XIV.2.C that there is a PL homemorphism h_2 of R^3 onto itself such that $h_2h_1(S_1) = $ Bd $v_0v_1v_2v_3$ and $h_2h_1(D)$ is the face $v_1v_2v_3$.

Let x_0 be the barycenter of $v_1v_2v_3$ and av_0x_0b be a straight line segment with points as ordered so that the join of av_0x_0b and Bd $v_1v_2v_3$ lies in $h_2h_1(U) \cup$ Bd $v_1v_2v_3$ and intersect $h_2h_1(S_2)$ only in $v_1v_2v_3$. Then a and b are near v_0, x_0, respectively. Let h_3 be a homeomorphism that is the identity outside this join and is defined on the join as follows. The homeomorphism h_3 takes av_0 linearly onto ax_0 and v_0b linearly to x_0b. If $p \in$ Bd $v_1v_2v_3$ and $q \in av_0x_0b$, h_3 takes pq linearly onto the segment from p to $h_3(a)$. Note that h_3 is a push from v_0 to x_0.

The required homeomorphism h is defined by $h_1^{-1}h_2^{-1}h_3h_2h_1$. $\square$

The following result follows from Theorem XIV.3.A.

THEOREM XIV.3.B. *Suppose that S_1, S_2 are tetrahedral 2-spheres in R^3 such that $S_1 \cap S_2$ is a disk D. Then $(S_1 \cup S_2) -$ Int D is tetrahedral.*

PROOF. Let $S_3 = S_1 \cup S_2 -$ Int D and U be an open set containing $S_3 \cup$ Int S_3. The proof is in two cases.

Case where Int $D \subset$ Int S_3. In this case, $S_1 \cup$ Int $S_1 \subset S_3 \cup$ Int $S_3 \subset U$ and $S_2 \cap$ Int $S_1 = \varnothing$. It follows from Theorem XIV.3.A that there is a PL homeomorphism h_1 of R^3 onto itself such that $h_1(S_3) = S_2$ and h_1 is fixed outside U. Since S_2 is tetrahedral, there is a PL homeomorphism h_2 of R^3 onto itself such that $h_2(S_2)$ is the boundary of a tetrahedron and h_2 is fixed outside U. In this case the required homeomorphism is h_2h_1.

Case where Int $D \subset$ Ext S_3. Since D does not separate R^3 (Theorem VIII.5.D) one of $S_1 - D$, $S_2 - D$ (say $S_2 - D$) is accessible from the unbounded component of $R^3 - (S_1 \cup S_2)$. Then $S_3 \not\subset S_1 \cup$ Int S_1, and it follows from Theorem XIV.3.A that there is a PL homeomorphism h_1 of R^3 onto itself such that $h_1(S_3) = S_2$. Since S_2 is tetrahedral there is a PL homeomorphism h_2 of R^3 onto itself and a 3-simplex Δ_1^3 such that $h_2(S_2) = $ Bd Δ_1^3. The PL homeomorphism h_2h_1 sends S_3 into Bd Δ_1^3, but we are not sure that h_2h_1 is fixed outside U.

Let Δ_2^3 be a 3-simplex in Int Δ_1^3 such that $(h_2h_1)^{-1}$ is linear on Δ_2^3 and h_3 be a PL homeomorphism of R^3 onto itself such that $h_3(\Delta_1^3) = \Delta_2^3$ and h_3 is fixed outside $h_2h_1(U)$. In this second case, the required homeomorphism is $(h_2h_1)^{-1}h_3(h_2h_1)$. $\square$

XIV.4. Special case of the PL Schoenflies theorem for R^4. Roughly speaking, we say that one polyhedral 2-sphere in R^3 is simpler than another if its horizontal cross sections are less complicated. One of the least complicated situations is the one in which each horizontal cross section is either a point (at the two extremities) or a simple closed curve. We treat that special case in this section. Perhaps some of the more subtle parts of the proof of the PL Schoenflies theore for R^4 occur in this special case since some of the proofs in the literature were questionable in the case. See [A_2] and [M_7].

For each polyhedral 2-sphere S in R^3 there is a triangulation of R^3 such that S is the union of some 2-simplexes of the triangulation. Suppose S is a polyhedral 2-sphere in R^3, B^3 is a large 3-cell whose interior contains S, and T is a triangulation of B^3 such that S is the union of certain 2-simplexes of T. Suppose B^3 is tilted so that no two vertices of T are in the same horizontal plane.

PROOF OF THE SPECIAL CASE OF THE PL SCHOENFLIES THEOREM FOR R^3. We suppose that no two vertices of T lie in the same horizontal plane and that each component of each horizontal cross section of S is either a point or a simple closed curve. At most, a finite number of the components are points, and it follows from 5.A of the Appendix that there are precisely two such points. These two points would be the highest and lowest points of S; and it could be shown that the other components are simple closed curves. Let $P_0, P_2, \ldots, P_{2n}$ be the horizontal planes through vertices of T with P_{2i-2} below P_{2i}. Let P_{2i-1} be a horizontal plane between P_{2i-2} and P_{2i}. Let S_i be the polyhedral 2-sphere which is the union of the part of S between P_i and P_{i+1}, the disk (or point) in P_i bounded by $P_i \cap S$, and the disk (or point) in P_{i+1} bounded by $P_{i+1} \cap S$. The special case of the PL Schoenflies theorem for R^3 will follow from the θ-curve Theorem XIV.3.B if it is shown that each S_i is tetrahedral.

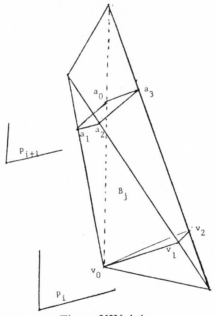

Figure XIV.4.A

It is relatively easy to see that S_0 is tetrahedral. Consider the intersection of $S_0 \cup \text{Int } S_0$ with the 3-simplexes of T that intersect $\text{Int } S_0$. Order these intersections $B_1, B_2, \ldots, B_k$ where $B_1 \cap P_1, B_2 \cap P_1, \ldots, B_k \cap P_1$ is a cellular shelling of $S_0 \cap P_1$. One proves that S_0 is tetrahedral by induction on the number of B's. Each Bd B_i is tetrahedral and by induction $\text{Bd}(B_2 \cup B_3 \cup \cdots \cup B_i)$ is also. Then $S_0 = \text{Bd}(B_1 \cup B_2 \cup \cdots \cup B_k)$ is tetrahedral.

For an arbitrary S_i, one also lets $B_1, B_2, \ldots, B_j$ be the intersections of $S_i \cup \text{Int } S_i$ and the 3-simplexes of T. We say that the B's can be *shelled* if they can be ordered $B_1, B_2, \ldots, B_j$ such that for each k, $B_k \cap (B_{k+1} \cup \cdots \cup B_j)$ is a disk. Again we use induction on the number of B's but it is not easy to pick an ordering that can be shelled. Note the difficulties in [$\mathbf{M}_7$]. One of the B's may look as shown in Figure XIV.4.A.

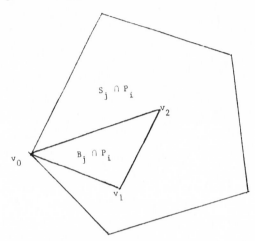

Figure XIV.4.B

Trouble below. Failing to shell the B's by looking at the top layer might lead one to examine the cellular subdivision of $S_i \cap P_i$ provided by the $P_i \cap B_j$'s. However, this does not provide a ready solution either as seen by Figure XIV.4.C. Here the triangular disk $v_0 v_1 v_2$ can be shelled off, but removing the associated B breaks the solid bounded by S_i into two pieces.

Trouble at the top. Let us examine some of the pitfalls in showing that the B's can be shelled from $S_i \cup \text{Int } S_i$. (Shelling is discussed in §XIV.6.) Assume P_i contains a vertex of T. Then P_{i+1} does not. It may be noted that the B's may be ordered so that $B_1 \cap P_{i+1}$, $B_2 \cap P_{i+1}, \ldots, B_k \cap P_{i+1}$ is a cellular shelling of $S_i \cap P_{i+1}$. However, if B_1 is removed, there is no assurance that the closure of $(S_i \cup \text{Int } S_i) - B_1$ is a 3-ball. If B_1 were selected because $B_1 \cap P_{i+1} = a_0 a_1 a_2 a_3$ as shown in Figure XIV.4.A shared the edge $a_0 a_1$ with $\text{Bd}(S_1 \cap P_{i+1})$, there is no assurance that $B_i \cap P_i = v_0 v_1 v_2$ shares any edge with $\text{Bd}(S_i \cap P_i)$. In fact, $B_i \cap P_i$ might appear as in Figure XIV.4.B.

A winning strategy. Once one recognizes that there is a problem, it is not difficult to devise a method for ordering the B's so that they can be shelled. First, ask if there is a face σ of one of the B's which does not lie on S_i but its boundary does. If there is such a σ, it is a spanning disk of $S_i \cup \text{Int } S_i$. It would have one edge on P_{i+1} and two edges between P_i and P_{i+1}. It would not have an edge on P_i if it were a triangular disk but would have one if it were trapezoidal. If there is such a face σ, then $S_i \cup \sigma$ is a θ-like 2-complex which is the union of three disks with a common boundary. Induction on the number of B's shows that two of the

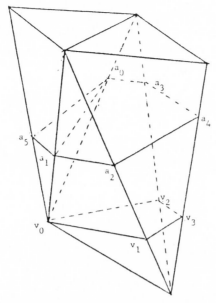

Figure XIV.4.C

2-spheres in $\dot{S}_i \cup \sigma$ is tetrahedral and it follows from Theorem XIV.3.B that S_i is also.

If there is no σ as mentioned above, then one can pick a B_j so that $B_j \cap P_i$ can be shelled from $P_i \cap S_i$ and show that B_j can be shelled from $S_i \cup \text{Int } S_i$. $\square$

XIV.5. General case of the PL Schoenflies theorem for R^3. Suppose in R^3 that S is a polyhedral 2-sphere with a triangulation $T(S)$. We suppose R^3 is tilted so that no two vertices of $T(S)$ lie in the same horizontal plane.

In the general case, we cannot suppose that each component of each horizontal cross section of S is a point or a polygonal simple closed curve. There may be some critical levels so that a horizontal cross section at each of these levels contains a bouquet of simple closed curves as shown in Figure XIV.5.A. If no

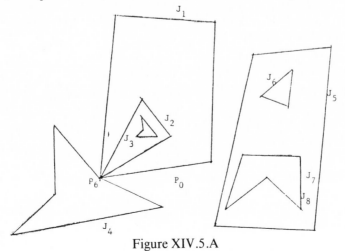

Figure XIV.5.A

horizontal cross section of S contains such a bouquet, then each component of each horizontal cross section of S is a point or a simple closed curve and we revert to a consideration of the special case which has already been treated.

Figure XIV.5.B shows an S with a horizontal cross section that resembles Figure XIV.5.A.a. It is a boundary of a 3-cell obtained by digging holes into a folded 3-ball and pushing out fingers.

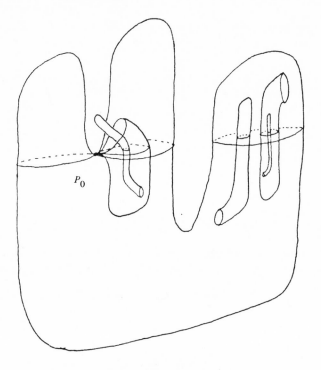

Figure XIV.5.B.a

Assume we do not have the special case and P_1, $P_2, \ldots, P_m$ are the critical horizontal planes such that $S \cap P_i$ contains a bouquet of simple closed curves. Let $n(P_i)$ be the number of simple closed curves in $S \cap P_i$ (not just the number in the bouquet). The $n(P_i)$ associated with Figure XIV.5.A.a is 8. Let $n(S) = \Sigma n(P_i)$. If S' is another polygonal 2-sphere in R^3 with a triangulation so that no two of its vertices lie in the same horizontal plane, we would possibly use a different set of critical planes in computing $n(S')$. A different tilt might change $n(S')$.

We use induction on $n(S) = \Sigma n(P_i)$ to show that S is tetrahedral. Assume that if S' is a polyhedral 2-sphere for which $n(S')$ is defined, then S' is tetrahedral if $n(S') < n(S)$.

Let P_0 be one of P_1, $P_2, \ldots, P_m$ and J_0 be a simple closed curve in $S \cap P_0$ such that the interior of the disk in P_0 bounded by J_0 contains no point of S. In Figure XIV.5.A, we could use any one of J_3, J_4, J_6, or J_8 as J_0. Let D_0 be the disk in P_0 bounded by J_0 and D_1, D_2 be the two disks in S bounded by J_0. It will follow from the θ curve Theorem XIV.3.B that $S = D_1 \cup D_2$ is tetrahedral if it is shown that each of $D_0 \cup D_1$, $D_0 \cup D_2$ is.

We finish the proof of the PL Schoenflies theorem for R^3 by showing that $S_1 = D_0 \cup D_1$ is tetrahedral. (Of course, $D_0 \cup D_2$ is also.) Let $T(S_1)$ be a triangulation of S_1 such that each vertex of $T(S_1)$ in P_0 is either a vertex of $T(S)$ or is on $J_0 = \mathrm{Bd}\, D_0$, no two vertices of $T(S_1)$ off P_0 lie in the same horizontal plane, and no simplex of $T(S_1)$ intersects both P_0 and another of the critical planes $P_1, P_2, \ldots, P_m$. Note that $T(S_1)$ gives D_0 a triangulation with no vertices on $\mathrm{Int}\, D_0$.

We are not in a position to compute $n(S_1)$ since P_0 contains several vertices of $T(S_1)$. We intend to adjust S_1 near P_0 to get a new 2-sphere S_1' so that $n(S_1')$ is defined and $n(S_1') < n(S)$. This will show that S_1' (and hence S_1) is tetrahedral. We suppose P_0 is horizontal and each part of J_0 is a limit part of the portion of D_1 below P_0.

Let $\sigma_1, \sigma_2, \ldots, \sigma_j$ be a shelling of the 2-simplexes of $T(S_1)$ in D_0 so that if J_0 contains a vertex p_0 of S, then $p_0 \in \sigma_j$. In any case, we let p_0, p_j, p_{j+1} be the vertices of σ_j. Since each vertex of each σ_i lies on J_0, each σ_i has a vertex p_i not on $\sigma_{i+1} \cup \sigma_{i+2} \cup \cdots \cup \sigma_j$. See Figure XIV.5.B.b.

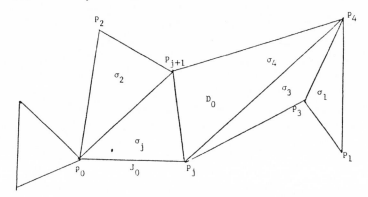

Figure XIV.5.B.b

Suppose that in a neighborhood of $J_0 - \{p_0\}$, D_1 is on or below P_0. We obtain the adjustment S_1' by pushing p_1 down slightly below P_0 (this changes each simplex in $T(S_1)$ that has p_1 as a vertex), then push p_2 down very slightly,..., and finally push p_{j+1} down very, very slightly. If p_0 is a limit point of $(S_1 \cap P_0) - D_0$, it is not pushed down; if it is not such a limit point, it is pushed down as a final step. This adjustment is made so that $n(S_1')$ is defined and if P' is a critical horizontal plane for S_1', then $S_1' \cap P' \subset S_1 \cap P'$. Since $n(P_0)$ as it applies to S_1' is less than $n(P_0)$ as it applies to S, $n(S_1') < n(S)$ and S_1' is tetrahedral. Since S_1' can be obtained from S_1 by a PL homeomorphism of R^3 onto itself, S_1 is also tetrahedral. $\square$

XIV.6. Shelling. Suppose T is a triangulation of a 3-manifold-with-boundary M^3. We say that T can be *shelled* if its 3-simplexes can be ordered $\Delta_1^3, \Delta_2^3, \ldots, \Delta_n^3$ so that $\Delta_i^3 \cap (\Delta_{i+1}^3 \cup \cdots \cup \Delta_n^3)$ is a disk for $1 < i < n$. There are many triangulations of PL balls that cannot be shelled. Theorem VI.6.A does not extend to 3-dimensions.

Suppose that instead of expressing M^3 as the union of 3-simplexes, we express it as the union of PL 3-balls such that if two of the 3-balls intersect, the intersection is a subset of the boundary of each. This decomposition of M^3 is called a *cellular subdivision*. We say that the balls can be *shelled* if they can be ordered $B_1^3, B_2^3, \ldots, B_n^3$ so that $B_i^3 \cap (B_{i+1}^3 \cup \cdots \cup B_n^3)$ is a disk. The collection of B's is called a cellular subdivision rather than a triangulation, and the shelling is called a *cellular shelling*. It follows from a proof rather than from the hypothesis that each $B_i \cup B_{i+1} \cup \cdots \cup B_n$ is a 3-ball. A favorite way of trying to show that a 3-manifold-with-boundary is a 3-ball is to show that it has a cellular subdivision that can be shelled. We used a version of this attack in proving the special case of the PL Schoenflies theorem for R^3. The shelling of a disk, on the other hand, is often used to replace one disk with a simpler one. Perhaps the difference in usage influenced the difference in definitions.

Some of the mysteries of 3-dimensional topology may be hidden by our lack of knowledge about shelling. Does each triangulation of a homotopy 3-ball have a subdivision that can be shelled? (A *homotopy 3-ball* is a compact connected simply-connected 3-manifold-with-boundary whose boundary is a 2-sphere.) We can use horizontal levels to get a shellable subdivision if the homotopy 3-ball lies in R^3. Is there an approach if it lies in some other PL 3-manifold? Some hope to get a counterexample to the Poincaré conjecture by constructing a homotopy 3-ball with a triangulation which has no shellable subdivision. Steve Armentrout has approached the Poincaré conjecture from this point of view.

XIV.6.A. *The house-with-two-rooms.* We first describe a 2-dimensional version of this house-with-two-rooms and then a 3-dimensional version. The two-dimensional version is shown in Figure XIV.6.A.a.

To build this house one starts with the boundary of a rectangular solid $a_1 b_1 c_1 d_1 a_3 b_3 c_3 d_3$ and adds a center section $a_2 b_2 c_2 d_2$ as shown. The house now has two rooms but no entrances.

To make an entrance to the bottom room, a hole $e_1 f_1 g_1 h_1$ is cut in the interior of $a_1 b_1 c_1 d_1$ and a hole $e_2 f_2 g_2 h_2$ cut in $a_2 b_2 c_2 d_2$ directly beneath it. A tunnel is run from $e_1 f_1 g_1 h_1$ to $e_2 f_2 g_2 h_2$. To make an entrance to the top room, holes $i_2 j_2 m_2 n_2$ and $i_3 j_3 m_3 n_3$ are cut in $a_2 b_2 c_2 d_2$ and $a_3 b_3 c_3 d_3$ as shown and a tunnel run between them.

The rooms are no longer simply-connected because of the column in each. Partitions $p_1 q_1 q_2 p_2$, $r_2 s_2 s_3 r_3$ are put in each as shown so that the rooms are now simply-connected. This completes the 2-dimensional house-with-two-rooms. Its complement is simply-connected.

Suppose that the house-with-two-rooms is made of cubical blocks laid face to face and one layer thick so that each vertex of a block is accessible from the outside. Although it may be difficult to visualize, the boundary of this solid house-with-two-rooms is a PL 2-sphere. Theorem XIV.1 implies that the solid is a 3-ball. However, the subdivision of it into blocks is not shellable. There is no place to start.

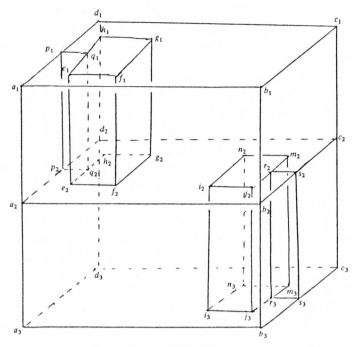

Figure XIV.6.A.a.

Before describing a triangulation of this solid house-with-two-rooms that cannot be shelled we put all the vertices in two classes. If two faces of a block are totally exposed and share an edge in common, we put all the vertices of the two faces in the second class. Any vertex never put in the second class (even as considering it on all blocks containing it) is put in the first class. By considering cases, one finds that each block has at least one vertex of the first class.

Let W be an ordering of the vertices of the block decomposition of the solid-house-with-two-rooms such that each vertex of the first class precedes each

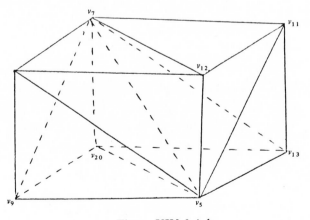

Figure XIV.6.A.b

of the second class. A face of a block is triangulated by drawing a 1-simplex from the first (in W) of its four vertices to the opposite vertex. A block itself is triangulated by coning from the first (in W) of its eight vertices. See Figure XIV.6.A.b. This gives a triangulation of the solid house-with-two-rooms that cannot be shelled. There is no place to start.

XIV.6.B. *Cube with knotted spanning 1-simplex.* A tame simple closed curve in R^3 (or S^3) is called *knotted* if it is not the boundary of any disk in R^3 (or S^3).

A spanning arc A of a tame 3-ball B^3 in R^3 is called *knotted* if there is an arc A' on Bd B^3 such that $A \cup A'$ is knotted. It is not important that $B^3 \subset R^3$ for we can require that $A \cup A'$ bounds no disk in B^3 rather than none in R^3.

Consider a tetrahedron $v_0v_1v_2v_3$ with a knotted polygonal spanning arc from v_0 to x_0, the barycenter of $v_1v_2v_3$. See Figure XIV.6.B.a. There is a curvilinear triangulation of $v_0v_1v_2v_3$ such that the spanning arc v_0x_0 is a 1-simplex of the triangulation. To get such a triangulation one could build a curvilinear cone in $v_0v_1v_2v_3$ so that v_0 is the apex of the cone, v_0x_0 is the axis, and a 2-simplex on $v_1v_2v_3$ having x_0 as center is the base. Subdivide the cone into three curvilinear tetrahedra with common edge v_0x_0. Build a buffer zone about the cone (similar to buffer zone in Theorem I.1.D and I.2.A), triangulate the complement of the buffered cone, and then subdivide the buffer being careful not to subdivide the cone further.

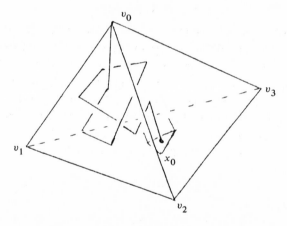

Figure XIV.6.B.a

It can be shown that having a knotted spanning 1-simplex is preserved through the stages of the shelling of a triangulated 3-ball. Hence, a triangulation of a 3-ball cannot be shelled if it contains a knotted spanning 1-simplex.

Rather than consider a curvilinear triangulation of a 3-ball, one can vary the 3-ball so as to get a rectilinear triangulation of it with a knotted spanning 1-simplex. Consider the cube with the plugged knotted hole shown in Figure XIV.6.B.b. The object is topologically a cube since the hole is plugged in the upper end with the small cube C'. The resulting topological cube is triangulated so that the edges of C' are 1-simplexes of the triangulation. This may be accomplished by first subdividing C into small cubes all the same size as C',

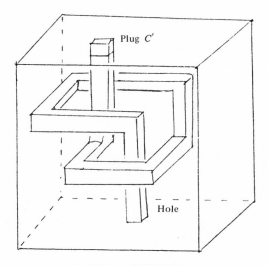

Figure XIV.6.B.b

subdividing each cube into two prisms, and then subdividing each prism into three tetrahedra—taking care that the resulting tetrahedra fit together right.

Consider a simple closed curve J made up of a spanning segment of C which is an edge of C' and an arc of Bd D. Then J is knotted and the triangulation cannot be shelled.

Closely associated with the notion of shelling is that of *collapsing*. Suppose T is a triangulation of a complex X. We call Δ_1 a face of Δ_2 if Δ_1, Δ_2 are elements of T with $\Delta_1 \subset$ Bd Δ_2. We call Δ_1 a *free* face of Δ_2 if $\Delta_1 \not\subset$ Bd Δ_3 if Δ_3 is any simplex of T other than Δ_2. We say that T *collapses* to T_0 if Δ_1 is a free face of Δ_2, and we form T_0 by removing Δ_1 and Δ_2 from T but keep other simplexes (including other faces of Δ_2). The associated X is reduced by (Int $\Delta_1 \cup$ Int Δ_2) if dimension Δ_1 is positive—by ($\Delta_1 \cup$ Int Δ_2) if dimension $\Delta_1 = 0$. The composition of a finite number of collapses is frequently designated by

$$T = T_1 \searrow T_2 \searrow \cdots \searrow T_n.$$

It is called a *simplicial collapse*.

The notions of a cellular shelling and simplicial collapsing generalizes to that of a *geometric collapse*. Suppose P_0 is a subpolyhedron of geometric complex P and $B_1, B_2, \ldots, B_n$ are PL balls (not necessarily all the same dimension) in P. We say that $P = B_1 \cup B_2 \cup \cdots \cup B_n \cup P_0$ *geometrically collapses* to P_0 if $B_i \cap (B_{i+1} \cup \cdots \cup B_n \cup P_0)$ is a ball on Bd B_i for $1 \leqslant 1 \leqslant n$.

Let T be a triangulation of a 3-manifold-with-boundary M^3, T' be the first barycentric subdivision of T, and T'' the second. The star in T'' of each vertex of T' is a PL 3-ball and can be thought of as either a ball around a vertex of T, a bar about the middle half of an edge of T, a plate about the center position of a 2-simplex, or a plug from the center of a 3-simplex of T. It can be shown that a simplicial collapsing

$$M^3 = T_1 \searrow T_2 \searrow \cdots \searrow T_n$$

of M^3 to a point T_n implies a cellular shelling of M^3 where $B_i \cup B_{i+1} \cup \cdots \cup B_n$ is union of 3-simplexes of T'' intersecting T_i. A related crucial result is that if a 3-manifold-with-boundary M geometrically collapses to a point, it has a cellular subdivision that shells. This makes such an M^3 a 3-ball.

To show that a triangulated 3-manifold-with-boundary has a subdivision that collapses to a point may be as difficult as to show that it has a subdivision that shells. In transferring from shelling to collapsing one may just be getting out of one sinkhold and into another. We look for inherent properties of the triangulated 3-manifold-with-boundary to show that some subdivision of it shells or collapses. Lying in R^3 is not such an inherent property.

Mary Ellen Rudin [R_3] gave an example of a rectilinear triangulation of a rectilinear tetrahedron that could not be shelled. Chillingsworth [C_8] showed that this triangulation could be collapsed by showing that any rectilinear triangulation of any convex 3-ball in R^3 could be collapsed.

CHAPTER XV

COVERING SPACES

XV.1. Examples and definitions. A *covering map* is a special map from one topological space to another. We give some examples.

The wrapping n times of one simple closed curve J_1 around another J_2 provides an example of a covering map. We say that J_1 is an n-sheeted covering of J_2, and the wrapping function is the associated projection map. In case J_1, J_2 are circles in the complex plane with equations $|z| = r_1$, $|z| = r_2$, respectively, we may give the equation of the projection map ρ as $\rho(r_1 e^{i\theta}) = r_2 e^{in\theta}$.

There is a 2-sheeted covering of the projective plane P^2 by the 2-sphere S^2 under which diametrically opposite points of S^2 go onto the same point of P^2.

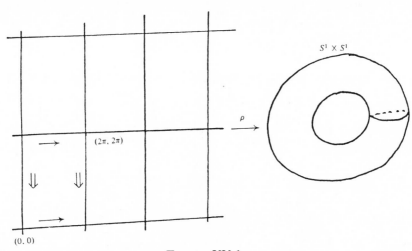

FIGURE XV.1

Figure XV.1 illustrates a familiar map of R^2 onto a torus $S^1 \times S^1$. It represents an infinite sheeted covering and could be given the equation $\rho(x, y) = (\sin x, \cos x) \times (\sin y, \cos y)$.

A space Y is an *n-sheeted covering* of X with a *projection map* ρ if there is an open covering $\{U_i\}$ of X such that for each $U_i \in \{U_i\}$ there is a collection of n

175

disjoint open subsets $V_{i,1}, V_{i,2}, \ldots, V_{i,n}$ of Y such that ρ takes each $V_{i,j}$ homeomorphically onto U_i and $\rho^{-1}(U_i) = \bigcup_{j=1}^n V_{i,j}$. It is intended that this definition permits n to be infinite.

The cartesian product of X with an n point discrete space provides a trivial example of an n-sheeted covering. However, we shall primarily be concerned with covering spaces which are connected.

This chapter will not contain an extended treatment of covering spaces. There are already many good treatments. Massey gives an easily understood introduction in [M_8, Chapter 5]. Instead, we only mention a few interesting things about covering spaces relevant to manifolds. We make use of Theorem XV.5.C in Chapter XVI. We made use of covering spaces in Chapter X, but it is not our purpose to rehash that here.

If one is given a connected space X, how does one find out if X has a connected 2-sheeted covering space? How does one build such a covering space if there is one? In §§XV.2, XV.4, and XV.5 we discuss methods of building covering spaces.

XV.2. Using splittings. Suppose $M^2 = X$ is a connected 2-manifold-without-boundary and J is a simple closed curve on M^2 such that J does not separate M^2. Suppose M^2 is split apart along J and M_1^2, M_2^2 are two copies of the split version of M^2 as shown in Figure XV.2. The figure shows the M_i^2's if J is the center line of an annulus in M^2; if J is the center line of a Moebius band, Bd M_i^2 has only one boundary component. An arc axb in M^2 that crosses J at x is split into two arcs ax' and $x''b$ in M_i^2. Build a 2-sheeted covering $\tilde{M}^2$ from M_1^2, M_2^2 by sewing their boundaries together as follows: attach x', x'' of M_1^2 to x'', x', respectively, of M_2^2. We have the following result.

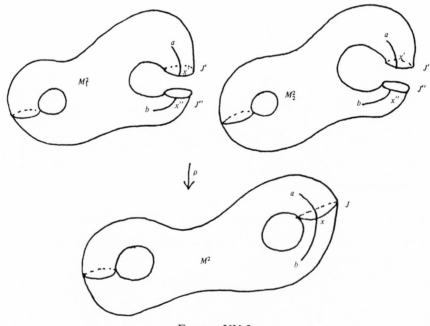

FIGURE XV.2.

THEOREM XV.2.A. *A connected 2-manifold-without-boundary M^2 has a connected 2-sheeted covering if M^2 contains a simple closed curve that does not separate M^2.* □

Since S^2 is the only compact connected 2-manifold that is separated by each simple closed curve in it, Theorem XV.2.A extends as follows.

THEOREM XV.2.B. *A compact connected 2-manifold has a connected 2-sheeted covering if it is not homeomorphic to S^2.* □

Our primary interest in covering spaces in this chapter is to apply them to 3-manifolds as follows.

THEOREM XV.2.C. *A connected PL 3-manifold M^3 (with or without boundary) has a connected 2-sheeted covering if there is a PL 2-manifold M^2 (perhaps with boundary) and a PL homeomorphism h of M^2 onto a connected closed subset of M^3 such that $M^3 - h(M^2)$ is connected and Bd $M^3 \cap h(M^2) = h(\text{Bd } M^2)$.* □

XV.3. Lifting paths. Before showing how to build covering spaces by using paths in §XV.4, we note some properties of paths. We recall that a path in X starting at $x \in X$ is a map of $[0, 1]$ into X such that $f(0) = x$. We say the path ends at $f(1)$. We call $f(0)$ its *initial point* and $f(1)$ its *terminal point*. Suppose $\tilde{X}$ is a covering space of X with the associated projection map ρ. We say that a *path f in X can be lifted* if for each $\tilde{x} \in \rho^{-1}f(0)$, there is a path $\tilde{f}$ in X started at $\tilde{x}$ such that $f = \rho\tilde{f}$.

THEOREM XV.3.A. *If $\tilde{X}$ is a covering space of X, then each path in X can be lifted.*

PROOF. It is convenient to prove a slightly stronger theorem—namely, that if $\tilde{x} \in \rho^{-1}f(0)$, there is a unique path $\tilde{f}$ in $\tilde{X}$ starting at $\tilde{x}$ such that $f = \rho\tilde{f}$. We note that for some points $t \in [0, 1]$ (those near $\tilde{x}$ and in a $V_{i,j}$ with $\tilde{x}$) there is a unique path $\tilde{f}_t$: $[0, t] \to \tilde{X}$ such that $\tilde{f}_t(0) = \tilde{x}$ and $f \mid [0, t] = \rho\tilde{f}_t$. One finds that the set of all points $t \in [0, 1]$ for which this is true is both open and closed in $[0, 1]$. Hence, each point t of $[0, 1]$ is such a point—in particular, 1 is. □

If f is a *map* of a set Y into X, we say that *f can be lifted* if for each $y \in Y$ and each point $\tilde{y} \in \rho^{-1}f(y)$ there is a map $\tilde{f}$: $Y \to \tilde{X}$ such that $\tilde{f}(y) = \tilde{y}$ and $f = \rho\tilde{f}$. Not each map of a circle into each X can be lifted. For example, if $\tilde{S}^1$ is a connected 2-sheeted covering of a circle S^1, then no homeomorphism of a simple closed curve into S^1 can be lifted. However, Theorem XV.3.A extends as follows.

THEOREM XV.3.B. *If $\tilde{X}$ is a covering space of X, then each map of a disk into X can be lifted.* □

If $\tilde{X}$ is a covering space of X with associated projection map ρ, we say that a *subset Y of X can be lifted* if for each point $y \in Y$ and each point $\tilde{y} \in \rho^{-1}(y)$, there is a map $\tilde{f}$: $Y \to \tilde{X}$ such that $\tilde{f}(y) = \tilde{y}$ and ρf is the identity map on Y. Theorem XV.3.B generalizes as follows.

THEOREM XV.3.C. *If $\tilde{X}$ is a covering space of X and Y is a connected locally arcwise connected subset of X such that each loop in Y can be shrunk to a point in X, then Y can be lifted.*   □

The following is another useful application of Theorem XV.3.B.

THEOREM XV.3.D. *Let $\tilde{X}$ be a covering space of X and f_1, f_2 be paths in X with the same initial and terminal points. Let $\tilde{f}_1, \tilde{f}_2$ be lifts of f_1, f_2 with the same initial points. Then $\tilde{f}_1, \tilde{f}_2$ have the same terminal point if the product $f_1 * f_2^{-1}$ can be shrunk to a point in X.*   □

If $x_0 \in X$ and $\tilde{X}$ is arcwise connected, Theorem XV.3.D implies that there are not more points in the lift of x_0 than there are elements in $\pi_1(X, x_0)$. If X is simply-connected, X is 1-sheeted. However, the simply-connected $\sin 1/x$ circle shown in Figure XV.3.D shows that this observation does not hold if we do not insist that $\tilde{X}$ be arcwise connected. In fact, if $x \in X$ and $\tilde{X}$ is arcwise connected, there are no more points in the lift of x than there are homotopy equivalence classes of paths from x_0 to x in X. (Recall that paths f_1, f_2 with the same terminal point belong to the same homotopy equivalence class if the product loop $f_1 * f_2^{-1}$ can be shrunk to a point in X.) This observation is relevant to the method we shall use in the next section to build covering spaces. Before proceeding there, we make a further observation.

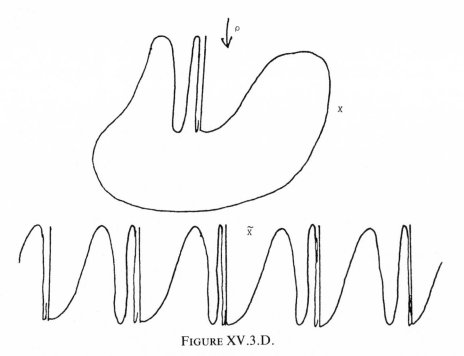

FIGURE XV.3.D.

THEOREM XV.3.E. *Suppose $\tilde{X}$ is a universal covering of a connected, 0-ULC, 1-ULC space K. Each point of X lifts into the same number of points as $\pi_1(X)$ has elements.*   □

XV.4. Using paths. We now discuss a method for building a covering space of a connected, 0-ULC, 1-ULC space X. Select a point $x_0 \in X$ to serve as a base point for $\pi_1(X, x_0)$. We shall build a connected covering space $\tilde{X}$ of X such that for each point $x \in X$, there is a 1-1 correspondence between the points of $\rho^{-1}(x)$ and homotopy equivalence class of paths from x_0 to x in X.

We shall take the abstract point of view and define a "point" of $\tilde{X}$ to be a homotopy equivalence class of paths in X starting at x_0. Any two paths in the same equivalence class would have the same ends and two paths f_1, f_2 from x_0 to x would be in the same equivalence class if the product $f_1 * f_2^{-1}$ can be shrunk to a point in X. It remains for us to define a topology for our space $\tilde{X}$ and show that it is a covering space of X.

Let $\{f_{x,\alpha}\}$ be a homotopy equivalence class of paths in X from x_0 to x. It is a point of $\tilde{X}$ and the projection map $\rho: \tilde{X} \to X$ takes $\{f_{x,\alpha}\}$ to x. We define the topology of $\tilde{X}$ so that ρ takes small connected open sets in $\tilde{X}$ homeomorphically onto small open sets in X.

For each point $x \in X$, let U_x' be an open neighborhood of x in X such that each loop in U_x' can be shrunk to a point in X. Let U_x be an open neighborhood of x in X such that each pair of points in U_x can be joined by an arc in U_x'. One neighborhood $N(\{f_{x,\alpha}\}, U_x', U_x)$ is the set of all $\{f_{y,\alpha}\}$ such that $y \in U_x$ and there is a path g_{xy} in U_x from x to y, a path $f_{x,0} \in \{f_{x,\alpha}\}$ and a path $f_{y,0} \in \{f_{y,\alpha}\}$ such that the product loop $(f_{x,0}) * (g_{xy}) * (f_{y,0})^{-1}$ can be shrunk to a point in X. It may be shown that the collection of all such open neighborhoods provides $\tilde{X}$ with a basis of open sets that causes $\tilde{X}$ to be a connected covering space of X.

A connected covering space of X is called a *universal covering space* of X if it is a covering space of each connected covering space of X. It may be shown that the covering space that we defined by using homotopy equivalence classes of paths is a universal covering space. It is simply-connected. Any connected simply-connected covering space is universal.

THEOREM XV.4.A. *Each connected, locally arcwise connected, locally simply-connected space has a universal covering space.* $\square$

Theorem XV.4.A extends. Instead of supposing that X is 1-ULC we could have merely required that small simple closed curves in X shrink in X. We shall not pursue generalizations since we shall only be applying Theorem XV.4.A where X is a polyhedron.

Instead of breaking the paths in X from x_0 to x into homotopy equivalence classes, one might have used other criteria for putting them into equivalence classes. For example, we might have selected a subgroup N of $\pi_1(X, x_0)$ and decreed that the paths f_1, f_2 from x_0 to x in X were in the same equivalence class if and only if the product loop $f_1 * f_2^{-1}$ was in an element of N. Such a procedure might enable one to break the paths into fewer equivalence classes (each class larger) and get a covering space with fewer sheets.

THEOREM XV.4.B. *Suppose X is connected, 0-ULC, and 1-ULC. Let $x_0 \in X$ and N be a subgroup of $\pi_1(X, x_0)$. Then there is a connected covering space $\tilde{X}$ of X such that a loop f in X based at x_0 lifts into $\tilde{X}$ if and only if f lies in an element of N.* ☐

Instead of using homotopy to break the paths in X from x_0 to x into equivalence classes we might have used singular first homology instead. Instead of using homology with integer coefficients there is some advantage in using homology with coefficient group Z_2 as we did in Chapter VIII. This may reduce the number of sheets in the covering space. One might reduce the numbers of sheets still further by selecting a subgroup of the first homology group and deciding that two paths f_1, f_2 are in the same equivalence class if $f_1 * f_2^{-1}$ lies in an element of the subgroup. We employ these techniques in the next section.

XV.5. A 2-sheeted covering. In Chapter XVI we use the result that if M^3 is a compact connected PL 3-manifold-with-boundary with a boundary component that is not a 2-sphere, then M^3 has a connected 2-sheeted covering. This follows from Theorem VIII.1.E and the following result. In its statement $H_1(M^3, Z_2)$ denotes the first homology group of M^3 with coefficient group Z_2.

THEOREM XV.5.A. *Suppose M^3 is a compact connected PL 3-manifold-with-boundary such that $H_1(M^3, Z_2)$ is nontrivial. Then M^3 has a connected 2-sheeted covering with a PL projection.*

PROOF. Let $a_1, a_2, \ldots, a_n$ be a minimal set of generators for $H_1(M^3, Z_2)$. Each loop in X corresponds to an element of $H_1(M^3, Z_2)$. Following the procedure we described in the preceding §XV.4, we put the paths in X from x_0 to x into equivalence classes. We decree that paths f_1, f_2 belong to the same equivalence class if and only if the product loop $f_1 * f_2^{-1}$ lies in the subgroup generated by $a_2, a_3, \ldots, a_n$. This breaks the set of paths into precisely two equivalence classes and insures that M^3 has a connected 2-sheeted covering.

Since maps of simplexes into M^3 lifts, we can suppose that if T is a triangulation of M^3, then there is a triangulation $\tilde{T}$ of $\tilde{M}^3$ with exactly twice as many simplexes as T and the projection map is linear with respect to taking $\tilde{M}^3$ under $\tilde{T}$ onto M^3 under T. ☐

We could have approached the proof of Theorem XV.5.A by way of Theorem XV.2.C through the following result.

THEOREM XV.5.A. *Suppose M^3 is a compact connected PL 3-manifold-with-boundary such that $H_1(M^3, Z_2)$ is nontrivial. Then there is a connected PL 2-manifold-with-boundary M^2 in M^3 such that*

$M^2 \cap \mathrm{Bd}\, M^3 = \mathrm{Bd}\, M^2$ *and*

$M^3 - M^2$ *is connected.*

PROOF. Our first objective is to give M^3 a cellular subdivision W such that the union of any collection of 3-cells of the subdivision is a PL 3-manifold-with-boundary. To this end we let T be a triangulation of M^3 and T', T'' be its first and second baricentric subdivisions. The closed 3-balls in the cellular subdivision are the stars in T'' of the vertices of T'. We call them W's and note that they are the balls, bars, plates, and plugs of §XIV.6.

If two W's intersect each other, their intersection is a disk. The union of any subcollection of these disks is a 2-manifold unless three of the disks share a common edge. No four of the disks ever share such an edge.

Let $D_1, D_2, \ldots, D_n$ be an ordering of these disks and $W_1, W_2, \ldots, W_m$ be an ordering of the W's so that Int W_1, Int $W_1 \cup$ Int $D_1 \cup$ Int $W_2, \ldots,$ Int $W_1 \cup$ Int $D_1 \cup \cdots \cup$ Int $D_{m-1} \cup$ Int W_m is a sequence of open 3-balls. The 2-manifold promised by Theorem XV.5.B will be a subset of $\cup_{i=m}^{n} D_i$. We next develop a method to decide which of the D_i's lie in M^2.

Let $a_1, a_2, \ldots, a_j$ be a minimal set of generators of $H_1(M^3, Z_2)$.

Let x_i be a point in Int D_i and L_i be a PL simple closed curve in $(M^3 - \cup_{i=m}^{n} D_i) \cup \{x_i\}$ which pierces D_i at x_i. We put D_i in M^2 if L_i corresponds to an element of $H_1(M^3, Z_2)$ which involves a_1—otherwise we throw it away.

Not all of the D_i's were thrown away since there are some loops in M^3 corresponding to elements of $H_1(M^3, Z_2)$ involving a_1.

To show that M^2 is a 2-manifold with $M^2 \cap$ Bd $M^3 =$ Bd M^2, one uses the fact that if K is a θ curve in M^3 one of whose loops corresponds to an element of $H_1(M^3, Z_2)$ not in the subgroup generated by $a_2, a_3, \ldots, a_n$, then a second loop in K has this property while the third does not. We use this result to show that no three of the kept D_i's share an edge and that Bd $M^2 \subset$ Bd M^3.   □

By putting together Theorems VIII.5.E, XV.2.C, and XV.5.A we get the following result.

THEOREM XV.5.C. *Suppose M^3 is a compact connected PL 3-manifold-with-boundary. Then M^3 has a connected 2-sheeted covering if either some boundary component of* Bd M^3 *is not a 2-sphere or M^3 contains a PL 2-manifold-with-boundary M^2 such that $M^3 - M^2$ is connected and $M^2 \cap$ Bd $M^3 =$ Bd M^2.*   □

CHAPTER XVI

DEHN'S LEMMA

XVI.1. Singular disks. Suppose M^3 is a PL 3-manifold-without-boundary and d is a PL map of a 2-simplex Δ^2 into M^3. If d is a homeomorphism, we call $d(\Delta^2)$ a PL *disk D* and if d is not a homeomorphism, we call $d(\Delta^2) = D$ a PL *singular disk*. We shall be interested not only in the singular disk D itself but also in the map d.

The *singularity set* of the map d (denoted by $S(d)$) is the closure of the set of all points $p \in \Delta^2$ such that $\{p\} \neq d^{-1}d(p)$.

We prefer to consider a disk as a point set rather than a map. (We mention this since some authors define a disk to be a map.) Then D is a singular disk whose boundary is Bd $D = d(\text{Bd } \Delta^2)$ and whose interior is Int $D = d(\text{Int } \Delta^2)$. The *set of singularities of D* is $S(D) = d(S(d))$.

The singular disk D is partially independent of the map d. For example, if h is a homeomorphism of Δ^2 onto itself, we regard $dh(\Delta^2)$ as the same singular disk as $d(\Delta^2)$; it has the same boundary, same interior, same singularity set, and is the same as a point set. We do not claim d and dh have the same singularity set.

We shall be primarily interested in singular disks whose singularity sets have certain nice properties.

XVI.2. Nice singularities. We shall study a singular disk D embedded in M^3 in a special way. The set of singularities of the special map d is a 1-complex in Δ^2. Figure XVI.2 shows how the embedded singular disk D looks near $S(D)$.

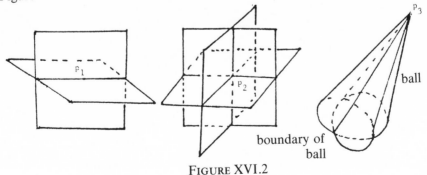

FIGURE XVI.2

183

We indicate in the left part of Figure XVI.2 that p_1 *lies on a double line*. There is a neighborhood U of p_1 in M^3 and a PL homeomorphism h of U onto R^3 such that $h(U \cap D)$ is the union of two perpendicular planes, $d^{-1}h^{-1}$ takes this union onto two disjoint open disks in Δ^2 and hd takes each of these open disks homeomorrphically onto one of the planes. Components of the union of all such p_i's are simple closed curves or open arcs. They are called *double lines*.

We indicate in the center part of Figure XVI.2 that p_2 is a *triple point*. Here $h(U \cap D)$ is the union of three mutually perpendicular planes, $d^{-1}h^{-1}$ takes this union onto three disjoint open disks in Δ^2, and hd takes each of these open disks homeomorphically onto one of the planes.

We indicate in the right part of Figure XVI.2 that p_3 is a *branch point*. A part of D near p_3 is the cone from the center of a 3-ball over a normal singular closed curve on the boundary of the ball. (We define a normal singular closed curve in the following paragraph.)

Consider a map j of a 1-sphere S^1 into a 2-sphere S^2. We want j to have only a finite number of singularities and each of them to be of order 2 i.e., if p is the image of a singular point, $j^{-1}(p)$ has precisely two points. Also, $j(S^1)$ crosses itself at such a point p in the sense that there is a neighborhood U of p in S^2 and a homeomorphism h of U onto R^2 such that $h(U \cap j(S^1))$ is the union of two straight lines intersecting at $h(p)$, $j^{-1}h^{-1}$ takes this union onto two disjoint open arcs in S^1, and hj takes each of these open arcs homeomorphically onto one of the straight lines. We call such a map j *normal* and call $j(S^1) = J$ a *normal singular closed curve* on S^2. We do not require that j be either PL or smooth, but it may simplify our thinking to regard it as "nice."

A PL singular disk D embedded in a PL 3-manifold M^3 so that its singularities consist of double lines, triple points, and branch points is called a *normal Dehn disk*. Note that none of the singularities of D lies on Bd D.

XVI.3. Splitting and resewing. Let us consider a favorite method of eliminating certain kinds of singularities from a normal Dehn disk. If the part of D near a double line (such as shown in the left part of Figure XVI.2) is replaced as shown in Figure XVI.3, we say that D was changed by *splitting along the double line and resewing*. This scissors-and-paste operation has been defined and discussed extensively in the literature (for example, Dehn [**D₂**, p. 150], Whitehead [**W₅**, p. 66], Papakyriakopoulos [**P₁**, pp. 4–6, **P₂**, pp. 284–286]). It is our purpose to use these cuts rather than to make a study of them here. We suppose that adjustments can be made as indicated in the figure.

We use these splitting and resewing operations to prove the following theorem.

THEOREM XVI.3.A. *Suppose D is a normal Dehn disk in a PL 3-manifold-without-boundary M^3 such that D contains no triple points or branch points. Then D can be changed to a nonsingular disk by splitting D along its double lines and resewing.*

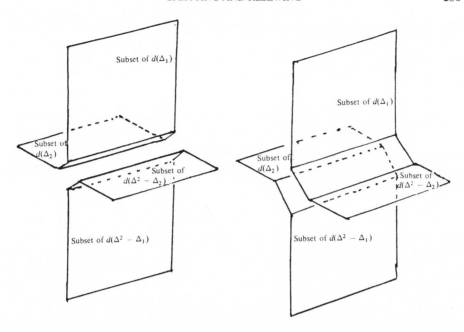

FIGURE XVI.3

PROOF. The simple closed curves in $S(D)$ are eliminated one at a time by splitting D along the double lines of the simple closed curve and resewing. In showing how to eliminate such a simple closed curve J in $S(D)$, we consider three cases.

Case 1. $d^{-1}(J)$ is the union of a pair of simple closed curves J_1, J_2 which bound a pair of disjoint disks Δ_1, Δ_2 in Δ^2. See Figure XVI.3.A.a. In this case we would adopt the sewing shown in the right part of Figure XVI.3 since this would trade disks, eliminate some singularities, and result in a simpler Dehn disk. If we used the sewing shown in the left part of Figure XVI.3, the resulting object would not be a Dehn disk but the union of a 2-sphere and a disk with a handle.

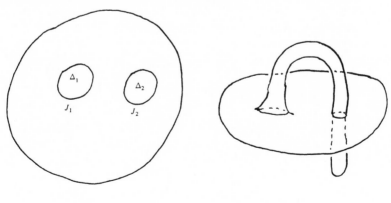

FIGURE XVI.3.A.a

Case 2. $d^{-1}(J)$ is a pair of concentric simple closed curves J_1, J_2 in the sense that J_1, J_2 bound disks Δ_1, Δ_2 in Δ^2 with $\Delta_1 \subset$ Int Δ_2. See Figure XVI.3.A.b. We are now confronted with a different situation. If we used the sewing suggested in the right part of Figure XVI.3, we would get the union of a Dehn disk and something obtained by sewing the two boundary components of the image of the annulus $\overline{\Delta_2 - \Delta_1}$ together. The left side of Figure XVI.3 provides a better procedure since it flips the image of the annulus $\Delta^2 -$ Int Δ_1, eliminates some singularities, and results in a Dehn disk with fewer singularities.

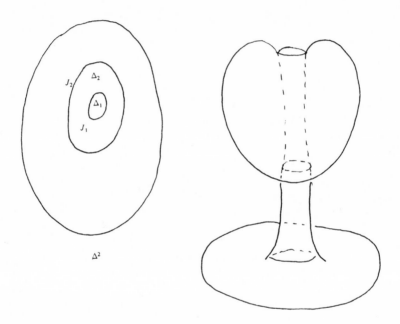

FIGURE XVI.3.A.b

Case 3. $d^{-1}(J)$ is one simple closed curve J_1. An investigation would show that in this case J is the center line of a solid Klein bottle, but we shall not use this fact. Rather, we consider the disk Δ_1 in Δ^2 bounded by J_1. One splits D along J and resews it as shown in Figure XVI.3.A.c being careful that the leaves which are subsets of $d(\Delta_1)$ are joined to those that are subsets of $d(\overline{\Delta^2 - \Delta_1})$. If one starts the splitting and resewing at a place of J_1 (or at two places of J) and circles J_1 (or goes halfway around J), the final sewing can be completed because leaves are joined in this prescribed fashion. We have essentially removed $d(\Delta_1)$ and sewn it in differently. If we had sewn leaves in the other fashion, we would have obtained the union of a Moebius band and a projective plane. $\square$

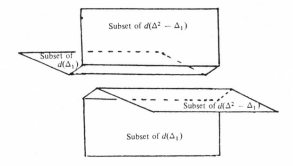

FIGURE XVI.3.A.c

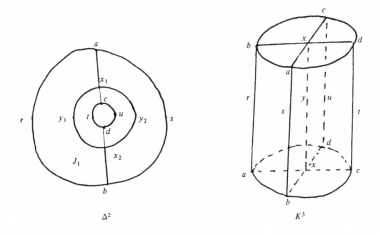

FIGURE XVI.3.A.d

The right side of Figure XVI.3.A.d shows a solid Klein bottle K^3 obtained by sewing the top of a solid cylinder onto the bottom in a nonorientable fashion. The simple closed curve *cudt* is the centerline of an annulus on Bd K^3. A pill box (not shown in Figure XVI.3.A.c) is sewn to K^3 along the annulus. Then *asbr* is the boundary of a normal Dehn disk D in the union of K^3 and the pill box. The left part of the figure shows a Δ^2 for Case 3.

XVI.4. Eliminating branch points. At a branch point, a normal Dehn disk is a cone over a normal singular curve. Each singular point of the curve determines a double line of points reaching to the branch point. The number of these double lines reaching to the branch point is called the *multiplicity of the branch point*. For example, the branch point p_3 shown in Figure XVI.2 is of multiplicity 3 while the cone over a figure eight would be of multiplicity 1.

A complicated normal singular closed curve on a 2-sphere can be gradually deformed to a simpler normal curve. For example, if a part of the normal curve shown in the left part of Figure XVI.4 were pushed up, there results the normal curve shown in the right part of Figure XVI.4$_1$ with fewer singularities. If this deformation is performed as one moves toward the branch point, one reduces the multiplicity of the branch point. Figure XVI.4$_2$ shows how the branch point in the right part of Figure XVI.2 might be simplified. Two of the double lines going to the branch point joined before they got there and became a single double line. The multiplicity of the branch point was reduced. Figure XVI.4$_3$ shows the singularity set before and after the deformation.

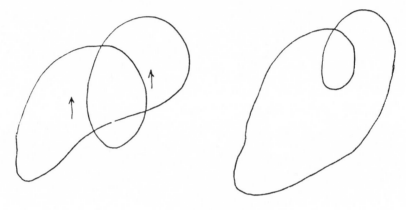

FIGURE XVI.4$_1$

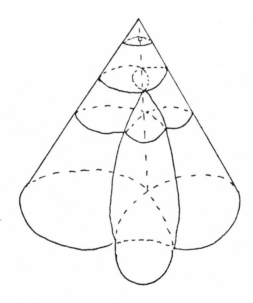

FIGURE XVI.4$_2$

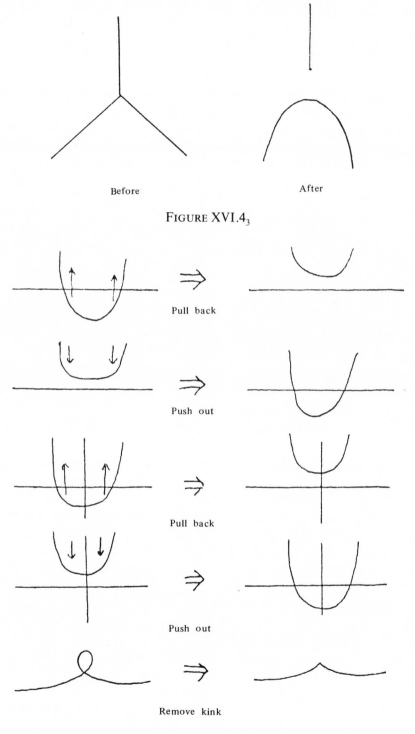

Before

After

FIGURE XVI.4₃

Pull back

Push out

Pull back

Push out

Remove kink

FIGURE XVI.4₄

Figure XVI.4$_4$ shows moves that can be used to simplify normal closed curves on a 2-sphere. The first is the pull-back operation discussed in the preceding paragraph. The second is the inverse of this move. When it is performed, a new double line is started but no new branch point is formed. The third and fourth moves represent pulls and pushes across vertices, and although these introduce triple points, they do not introduce new branch points. Pulling out a kink introduces a branch point of multiplicity one. Hence, by deforming the base of the cone at a branch point one can change the normal Dehn disk so that all its branch points have multiplicity one. (Had we exercised more care, by pushing the curve over the back side of the 2-sphere, we could have changed the normal disk without performing the kink-removing operation more than once. However, it was our immediate purpose to get all branch points to have multiplicity one rather than to avoid increasing their number.)

The proof of the following theorem follows either from the preceding two paragraphs or from Papakyriakopoulos' Lemma 3.1 on [**P$_1$**, p. 6]. Papakyriako-poulos' argument may be easier to understand than the one we suggested.

THEOREM XVI.4.A. *Suppose D is a normal Dehn disk in a PL 3-manifold M^3 and U is a neighborhood in M^3 of the union of the branch points of D. Then there is a normal Dehn disk $E = e(\Delta^2)$ in M^3 such that*

each branch point of E is of multiplicity 1,
$e = d$ *on* $\Delta^2 - d^{-1}(U)$, *and*
$ed^{-1}(U) \subset U$.   □

Suppose one has a branch point of multiplicity 1 as shown in the left part of Figure XVI.4.A. Then by modifying the map d only in a neighborhood in Δ^2 of $d^{-1}[pv]$, one can get the normal Dehn disk a piece of which is shown in the center part of Figure XVI.4.A. The inverse in Δ^2 of this center part is obtained by removing from Δ^2 a disk with center at $d^{-1}(p)$ whose boundary passes through the two points of $d^{-1}(v)$, then identifying the two points and finally plugging in the two holes with disks whose images are the small cones shown in Figure XVI.4.A. The points p, v, and s are branch points of the left, center, and right parts, respectively, of the figure. One can continue splitting along the double line from the branch point and perhaps past triple points (hence eliminating them) as shown in the right part of Figure XVI.4.A.

If one continues the splitting as far as possible, the only thing that blocks the splitting is to arrive at another branch point q. Let B^3 be a small polyhedral star-like ball centered at q and suppose the splitting is continued until the remaining part of the double line from q lies in Int B^3. The normal Dehn disk (which we still call D) intersects Bd B^3 in either one or two PL simple closed curves. (A careful analysis shows that it intersects in two simple closed curves J_1, J_2 so that $d^{-1}(J_1)$, $d^{-1}(J_2)$ are concentric on Δ^2, but we shall not use this fact.) If $D \cap$ Bd B^3 is one simple closed curve (which it is not), we obtain a new normal Dehn disk with less singularities by replacing $D \cap B^3$ by a disk in B^3. If

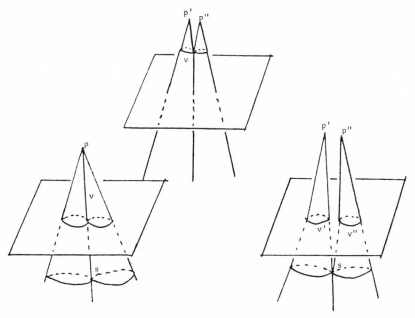

FIGURE XVI.4.A.

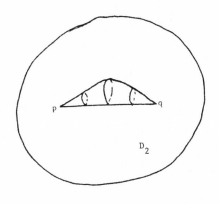

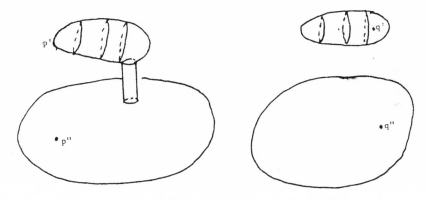

FIGURE XVI.4.B

$D \cap \mathrm{Bd}\, B^3$ is the union of two simple closed curves whose inverses in Δ^2 are not concentric, we would replace $D \cap B^3$ by the union of two disks in B^3. If $D \cap \mathrm{Bd}\, B^3$ is the union of two simple closed curves whose inverses in Δ^2 are concentric, we replace $D \cap B^3$ with an annulus in B^3. Had we been willing to discard a part of D, we could have capped over the curves rather than made an annular replacement. A disk with a roll feathered out at the ends is shown in the top part of Figure XVI.4.B. If one had followed a splitting with such a capping one would have obtained the right part of the figure rather than the disk shown in the left part.

We note that if each branch point is of multiplicity 1, then these branch points occur in pairs and can be eliminated two at a time.

THEOREM XVI.4.B. *If D is a normal Dehn disk in a PL 3-manifold-without-boundary M^3, then by modifying D near the branch points of D and splitting along double lines (possibly through triple points) and resewing, one can obtain a normal Dehn disk $E = e(\Delta^2)$ such that E has no branch points. For each neighborhood U in M^3 of $S(D)$ one can get such an E with*

$e = d$ *on* $\Delta^2 - d^{-1}(U)$,

$D - U = E - U,$ *and*

$ed^{-1}(U) \subset U.$ $\square$

A normal Dehn disk without branch points is called a *canonical Dehn disk*.

XVI.5. Trying to remove more singularities. After one has exhausted the methods used in §§XVI.3 and XVI.4, what can be done further to remove singularities from a canonical Dehn disk? One might try splitting along double lines, but there is no obvious starting point (such as there is when one has a branch point). One might try to start at an interior point of a double line by cutting out small wedges as shown in the left part of Figure XVI.5 and resewing as in the right part. The disks *abcd* and *aecf* were removed from the vertical and horizontal leaves, and the four disks *abf*, *cbf*, *ade*, *cde* were added. (Our figure does not show *ade* and *cde*.) The points *a* and *c* become branch points.

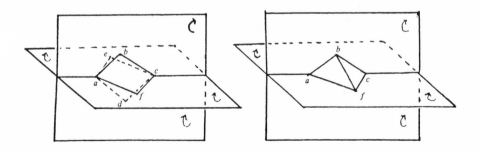

FIGURE XVI.5

There are several problems that arise in connection with this approach. One does not know which pair of adjacent leaves to connect. No matter which he chooses, he will no longer get a singular disk but rather a disk with a handle (perhaps nonorientable). There are now two branch points. One might hope that if one picked the right pair of adjacent leaves to join and then split out and resewed some double line joining the branch points, then one would recover a Dehn disk; this is difficult to prove. Could one pick a good start? (Splitting and resewing past triple points changes double lines.) (We suggest in §XVI.6 an approach that sometimes meets with success and in §XVI.8 one that always works.)

The arrows in Figure XVI.5 indicate orientation. If one is willing to introduce an orientable handle, then the choice of leaves to be joined is determined by orientation. For example, in Figure XVI.5, if the leaves are joined top to front and bottom to back, the original orientation is preserved and there results a disk with an orientable handle.

If one seeks a 2-manifold-with-boundary rather than a disk, one can use the methods suggested by Figure XVI.5. The resulting disk with handle can be split and resewn along the double line joining the branch points as suggested in §XVI.4. By performing the operation a finite number of times, one obtains the following result.

THEOREM XVI.5. *Suppose D is a normal Dehn disk in a PL 3-manifold-without-boundary M^3. Then by cut and paste along $S(D)$ one can change D to a connected nonsingular orientable 2-manifold-with-boundary F such that* Bd D = Bd F. *For each neighborhood U in M^3 of $S(D)$ one can pick such an F so that $D - U = F - U$.* □

Cut and paste. Let us consider what we mean by saying that one object can be obtained from a normal Dehn disk D by cut and paste along $S(D)$. Suppose D is changed to D' by splitting and resewing along each double line of D. There is no reason to expect that D' is a 2-manifold near the branch points and triple points of D. Consider a small PL 3-ball about each such branch point or triple point and replace the part of D' in the 3-ball with a 2-manifold-with-boundary so that the resulting D'' is a PL 2-manifold. We do not place restriction on this replacement, but it usually results from capping holes or sewing in annuli. Near double lines, the only operation permitted (unless it is allowed because the point is also near a vertex) is the splitting and resewing indicated by Figure XVI.3. No long tubes run along the 1-dimensional part of $S(D)$. (In the proof of Theorem XVI.6.B we shall use small annuli near triple points.) If we throw away any part of D not near $S(D)$, we say that we used *cut, paste* and *discard*.

XVI.6. Thickening disks. Suppose h is a PL homeomorphism of $\Delta^2 \times [-1, 1]$ into M^3. We call $h(\Delta^2 \times [-1, 1])$ a *thickening* of the PL disk $h(\Delta^2 \times 0)$. The $h(x \times [-1, 1])$'s are called fibers. It is convenient to think of the fibers as straight and perpendicular to $h(\Delta^2 \times 0)$, but this is not required by the definition.

Figure XVI.6 shows two intersecting disks and thickenings of these disks. The thickenings resemble slabs with the disks as center-slices. The fibers in the shown figure are straight and perpendicular to planar disks, but in general they need not be. The figure gives a PL view of how the fibers fit.

If D is a canonical Dehn disk in a PL 3-manifold M^3, a thickening $M^3(D)$ of D in M^3 is a 3-manifold-with-boundary in M^3 containing D that satisfies certain conditions. Near the double lines of D it resembles Figure XVI.6. At the triple points of D it resembles the union of three slabs. Elsewhere it locally resembles the thickening of a disk. Also, $D \cap \text{Bd}(M^3(D)) = \text{Bd } D$. For those acquainted with regular neighborhoods we would say that $M^3(D)$ is a regular neighborhood of D in $M^3(D)$. A *regular neighborhood* of a polyhedron P in an n-manifold-with-boundary is a submanifold-with-boundary W^n such that W^n collapses to P. Some authors (but not all) require that some open set in the original n-manifold-with-boundary contain P and lie in the regular neighborhood. A convenient way to get a regular neighborhood W^n of P in W is to let T be a triangulation of W so that P is a subcomplex, T'' be the second barycentric subdivision of T, and W^n be the union of the simplexes of T'' intersecting P. Regular neighborhoods of P are not unique, as there are others.

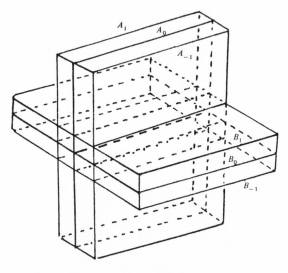

FIGURE XVI.6

It is clear that D can be thickened at places where it looks as nice as that shown in Figure XVI.2, but we realize that we are in a PL 3-manifold M^3 rather than in R^3 and D has some bends. Locally we could thicken parts of D, but there is the question as to how these pieces fit together. It is a valid area of study to show that things work out as pictured. The study would be facilitated by the fact that we are in the PL category, and we would not have some of the problems in making collars fit that we encountered in the proof of Theorem V.4.C. The technique of thickening canonical Dehn disks has been defined and studied extensively (for example, Johansson [$\mathbf{J_1}$, pp. 318–320], and [$\mathbf{J_2}$, p. 662]; Papakyriakopoulos [$\mathbf{P_1}$, pp. 8–9]; Shapiro and Whitehead [$\mathbf{SW_1}$]). These thickened disks have been called both prismatic neighborhoods and regular neighborhoods. It is our purpose to use these thickened disks rather than to make a detailed study of them here. We suppose that things are nice, as indicated.

Let us consider the case where Bd D bounds a disk E_1 on Bd $M^3(D)$ and no fiber has both ends in E_1. Then the disk E_1 can be used to construct a disk E such that Bd $D =$ Bd E and E lies in the union of D and a neighborhood in $M^3(D)$ of the singularity set of D. For example, if neither of the disks A_1 or A_{-1} of Figure XVI.6 were in E_1, we discard A_0 from D in building E. If A_{-1} and B_1 were in E_1, then we would split along the double line and sew A_0 to B_0 as indicated in the right part of Figure XVI.3. (Ignore the labels on Figure XVI.3 which were used for another purpose.) The disk E that we finally obtain is a projection of the disk E_1.

THEOREM XVI.6.A. *Suppose D is a canonical Dehn disk in a PL 3-manifold-without-boundary M^3 and $M^3(D)$ is a manifold-with-boundary obtained by thickening D. If* Bd $M^3(D)$ *contains a disk E_1 such that*

Bd $E_1 =$ Bd D, *and*

E_1 *does not contain both end points of any fiber,*

then D can be changed to a nonsingular disk E by cut, paste, and discard such that Bd $D =$ Bd E. *For each neighborhood U in M^3 of $S(D)$ one can get such an E so that $E \subset D \cup U$.* □

Theorem XVI.6.A lacks some elegance in that we used cut, paste, and discard rather than cut and paste alone and therefore had $E \subset D \cup U$ rather than $E - U = D - U$. We shall get this stronger result later when we prove a version of Dehn's lemma but for now recover it in the special case that $M^3(D)$ satisfies an additional condition.

THEOREM XVI.6.B. *Suppose D is a normal Dehn disk in a PL 3-manifold-with-boundary, $M^3(D)$ is a manifold-with-boundary obtained by thickening D and $M^3(D)$ has no connected 2-sheeted covering. Then by splitting and resewing along double lines of $S(D)$ and adjusting near triple points and branch points, one can use cut and paste (without discard) to change D to a nonsingular disk E such that* Bd $D =$ Bd E. *For each neighborhood U in M^3 of $S(D)$ one can get such an E so that $D - U = E - U$.*

PROOF. It follows from Theorem XV.5.A that the first homology of $M^3(D)$ of mod 2 (that is, $H_1(M^3(D), Z_2)$) is trivial. It follows from Theorem VIII.1.E that each boundary component of Bd $M^3(D)$ is a 2-sphere.

We show that if P is a polygon in Int $M^3(D)$ that is in general position with respect to D, then P intersects D in an even number of places. Since P bounds, it follows from Theorem VIII.1.C that it bounds a 2-manifold M_1^2. Also, as suggested in Theorem XVI.3.A we can change D to M_2^2—a 2-manifold with Bd D as a boundary. Suppose M_1^2 and M_2^2 are in general position and $M_1^2 \cap M_2^2$ is the union of disjoint arcs and simple closed curves. Since each arc has two ends, $P \cap M^2$ is even; hence $P \cap D$ is even.

Let $S_0, S_1, \ldots, S_n$ be the components of Bd $M^3(D)$ and S_0 be the one which contains Bd D. Let E_1, E_2 be the two disks in S_0 bounded by Bd D. We wish to put the elements of $S_1, S_2, \ldots, S_n$ into equivalence classes—one class associated with E_1 and the other associated with E_2.

Let A_i be a polygonal arc from E_1 to S_i in $M^3(D)$ so that $A_i \cap (\text{Bd } D \cup S(D)) = \varnothing$ and A_i pierces D at each point at which it intersects D. Put S_i in the equivalence class associated with E_1 if $A_i \cap D$ has an even number of points and in the class associated with E_2 if it has an odd number. The result about the even

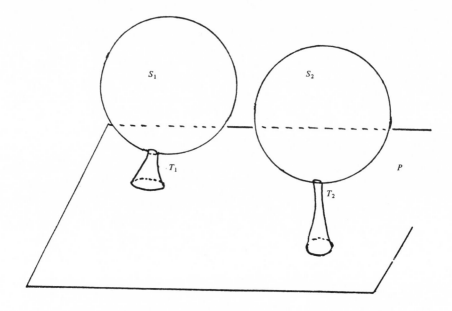

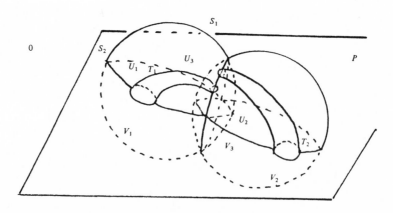

FIGURE XVI.6.C$_1$

number of piercing points of a polygon P with D shows that the placing of the S_i's into the equivalence classes is independent of the choices of the A_i's. If both ends of some fiber lie on Bd $M^3(D)$, then exactly one end lies in the union of E_1 and the S_i's associated with E_1 while the other end lies in the union of E_2 and the S_i's associated with E_2.

We shall use the union of E_1 and the S_i's associated with it to build an E. Just as one used cut, paste, and discard to get an E onto which E_1 projected in the proof of Theorem XVI.6.A, one can use cut and paste without discard along $S(D)$ to change D to the union of a disk E_0 and a finite collection of mutually disjoint 2-spheres $\{K_j\}$ such that E_1 projects onto E_0 and each S_i associated with E_1 projects onto some K_j. One can change the union of a disk and a 2-sphere to a disk by cutting a hole in the disk and a hole in the 2-sphere and joining the holes with a tube. Such operations can be done in neighborhoods of the triple points and branch points of $S(D)$ so as to convert the union of E_0 and the K_j's to one

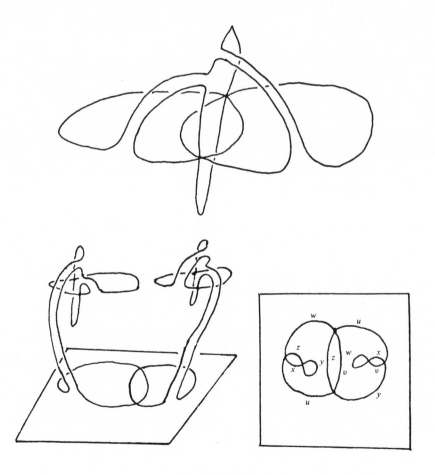

FIGURE XVI.6.C$_2$

disk E such that E is obtained from D by splitting and resewing along the double lines of $S(D)$ and adjusting near the triple points and branch points. Note that D was changed to E by cut and paste without discard.   □

EXAMPLE XVI.6.C. We now give an example of a Dehn disk D embedded in R^3 so as to satisfy the hypothesis of Theorem XVI.6.B. The example is shown schematically in Figure XVI.6.C$_1$ as the union of a planar disk P with two holes, two round 2-spheres S_1, S_2 each with a hole, and tubes T_1, T_2 joining the boundaries of the appropriate holes. In the actual example X_1, S_2 are pushed over so that they intersect P in circles as shown in the lower part of Figure XVI.6.C$_1$. The tube T_1 that once ran straight up to S_1 is now curved and bisected by S_2 while the tube T_2 that once ran straight up to S_2 is bisected by S_1. Before the tubes T_1, T_2 are inserted, the planar disk with X_1 and S_2 divides R^3 into an outside O, three top components U_1, U_2, U_3 as shown and three lower components V_1, V_2, V_3. Tubing connects O to U_1 and U_2 while it connects U^3 to V_1 and V_2. The technique of the proof of Theorem XVI.6.B shows how the example might be changed to a nonsingular disk by cut and paste. An investigation would show that one tubing operation would be required to prevent discard.

The upper part of Figure XVI.6.C$_2$ shows the set of singularities of the disk shown in Figure XVI.6.C$_1$. The lower part of Figure XVI.6.C$_2$ shows two views of the singular set of the map. Johansson has an interesting example in [J_1].

XVI.7. History of Dehn's lemma. In 1910 [D_2], Max Dehn first presented the lemma with a "proof"; however, in 1929 in [K_4], H. Kneser discovered a serious gap in the proof given by Dehn. At that time, Kneser could neither come up with a counterexample nor supply the proof necessary to fill the gap. The lemma remained a challenge for several decades. The lemma may be stated as follows.

LEMMA XVI.7 (DEHN'S LEMMA). *If D is a PL singular disk in a 3-manifold-without-boundary M^3 such that $S(D) \cap$ Bd $D = \varnothing$, then there exists a nonsingular PL disk, D_0, such that $D_0 \subset M^3$ and Bd $D_0 =$ Bd D.*   □

In 1938 in [J_2], I. Johansson proved that if Dehn's lemma holds for all orientable 3-manifolds, then it holds for all nonorientable ones. C. D. Papakyriakopoulos in 1957 in [P_1], proved that the lemma held for all orientable 3-manifolds, thus completing a proof of the lemma. In his proof, he also made use of another paper by I. Johansson on singular surfaces [J_1]. Also appearing in 1957, independent of Papakyriakopoulos' proof, was a proof by T. Homma [H_9] of Dehn's lemma in S^3.

Since Papakyriakopoulos' proof of the lemma, there have been several generalizations and simplifications. In 1958 A. Shapiro and J. H. C. Whitehead [SW_1] presented a proof of Dehn's lemma and an extension of it. Their proof of the lemma is a simplification of Papakyriakopoulos' proof.

An extension and strengthening of Dehn's lemma was presented in 1965 in [H_8] by D. W. Henderson; in his proof, he also used the technique used by Shapiro and Whitehead. Henderson obtained Theorem XVI.8.A as stated in the next section. Chapter XVIII mentions some other extensions of the lemma in which some of the PL restrictions and singularity restrictions on D are dropped. However, we still do not know the answer to the following question.

Question. Suppose f is a map of a disk D into a 3-manifold-without-boundary M^3 such that f is a homeomorphism on Bd D and $f(\text{Bd } D) \cap f(\text{Int } D) = \varnothing$. Does $f(\text{Bd } D)$ bound some disk E in M^3? The question asks if a simple closed curve bounds a disk if it can be shrunk to a point in its own complement. Bing has shown that the answer is yes if $f(\text{Bd } D)$ is PL.

XVI.8. A strong version of Dehn's Lemma. In this section we prove the following version of Dehn's Lemma.

THEOREM XVI.8.A. *Each normal Dehn disk D in any PL 3-manifold-without-boundary M^3 can be changed to a nonsingular disk E by cut and paste without discard along $S(D)$.*

PROOF. If the theorem were not true, there would be a counterexample. In fact, it follows from Theorem XVI.4.B that there would be a counterexample without any branch point.

For each Dehn disk $D = d(\Delta^2)$ without branch points let $n(d)$ be the number of 1-simplexes in a minimal curvilinear triangulation of $S(d)$. We prove Theorem XVI.8.A by induction on $n(d)$.

Assume Theorem XVI.8.A is false. Then there is a canonical Dehn disk $D_0 = d_0(\Delta^2)$ in a PL 3-manifold-without-boundary M_0^3 such that D_0 cannot be changed to a nonsingular disk by cut and paste (without discard) in a thickening $M^3(D_0)$ of D, but if $D = d(\Delta^2)$ is a canonical Dehn disk in a PL 3-manifold-without-boundary M^3, then D can be changed to a nonsingular disk by cut and paste (without discard) in $M^3(D)$ if $n(d) < n(d_0)$. We prove the theorem by showing that there is no D_0.

Let $M^3(D_0)$ be a 3-manifold-with-boundary obtained by thickening D_0 in M_0^3. Since D_0 cannot be changed to a nonsingular disk by good cut and paste (without discard), it follows from Theorem XVI.6.B that $M^3(D_0)$ has a connected 2-sheeted covering—i.e., there is a connected PL 3-manifold-with-boundary $\tilde{M}^3(D_0)$ and a projection map $\rho: \tilde{M}^3(D_0) \to M^3(D_0)$ such that for each point $p \in M^3(D_0)$ there is a neighborhood U of p such that $\rho^{-1}(U)$ is the union of two disjoint open subsets $\tilde{U}_1$, $\tilde{U}_2$ of $\tilde{M}^3(D_0)$ such that ρ takes each of $\tilde{U}_1$, $\tilde{U}_2$ homeomorphically onto U.

Since a map of a disk can be lifted, there are maps d_1, d_2 of Δ^2 into $\tilde{M}^3(D_0)$ such that each $\rho d_i = d_0$ and $\rho^{-1}d_0(\Delta^2) = d_1(\Delta^2) \cup d_2(\Delta^2)$. Since $M^3(D_0)$ is a thickening of $d_0(\Delta^2)$, $\tilde{M}^3(D_0)$ is a thickening of $d_1(\Delta^2) \cup d_2(\Delta^2)$. Since $\tilde{M}^3(D_0)$ is connected, so is $d_1(\Delta^2) \cup d_2(\Delta^2)$. In fact, $d_1(\Delta^2) \cap d_2(\Delta^2)$ contains an arc A such that $\rho(A)$ lies on a double line of $d_0(\Delta^2)$, and $d_1^{-1}(A) \not\subset S(d_1)$ but $d_1^{-1}(A) \subset S(d_0)$. Hence $S(d_1)$ is a proper subset of $S(d_0)$. Since some open 1-simplex of $S(d_0)$ is missing from $S(d_1)$, $n(d_1) < n(d_0)$. It may be shown that $d_1(\Delta^2)$ is a canonical Dehn disk. Hence it follows by induction on $n(d)$ that $d_1(\Delta^2)$ can be changed by good cut and paste (without discard) to a nonsingular disk $E_0 = e_0(\Delta^2)$ in $\tilde{M}^3(D_0)$.

We now show that $\rho(E_0) = \rho e_0(\Delta^2)$ is a canonical Dehn disk in $M^3(D_0)$. Along double lines of D_0 which do not lift to double lines of $d_1(\Delta^2)$, $\rho(E_0)$ locally resembles D_0. Along double lines of D_0 which lift to double lines of $d_1(\Delta^2)$, $d_1(\Delta^2)$ was split on the lifted double line to obtain E_0 and ρ acts as a homeomorphism near the lifted double line. We now turn our attention to a triple point p of D_0. If $\rho^{-1}(p) \cap d_1(\Delta^2)$ contains only one point, this point is a triple point of $d_1(\Delta^2)$, $d_1(\Delta^2)$ has already been adjusted near this point to form E_0, and $\rho(E_0)$ looks like a 2-manifold near p. If $\rho^{-1}(p) \cap d_1(\Delta^2)$ contains two points, one of these points lies on a double line of $d_1(\Delta^2)$ while the other one lies on a nonsingular point of $d_1(\Delta^2)$. Then near p, $\rho(E_0)$ has two double lines L_1, L_2 as shown in Figure XVI.8.A. Hence $\rho e_0(\Delta^2) = \rho(E_0)$ is a canonical Dehn disk.

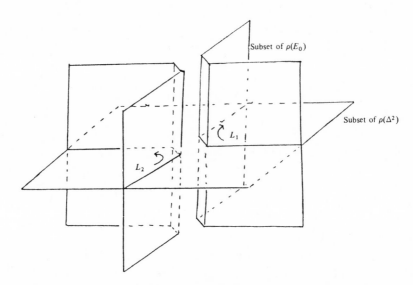

FIGURE XVI.8.A

Since ρ is a projection of a two-sheeted covering and E_0 is a nonsingular disk, $\rho(E_0) = \rho e_0(\Delta^2)$ has no triple points. Each component of $S(\rho(E_0))$ is a simple closed curve. Hence, it follows from Theorem XVI.3.A that we can change $\rho(E_0)$ to a nonsingular disk E by splitting $\rho(E_0)$ along its double lines and resewing. Then E is obtained from D_0 by good cut and paste along $S(D_0)$. This shows that there is no D_0 as supposed and Theorem XVI.8.A is established by induction on $n(d)$.

A summary of how we obtained E is suggested by

$$D_0 \to d_1(\Delta^2) \to E_0 \to \rho(E_0) \to E.$$

Here D_0 is a normal Dehn disk, $d_1(\Delta^2)$ is a lift of D_0 that is simpler than D_0, E_0 is a nonsingular disk obtained from $d_1(\Delta^2)$ by cut and paste without discard, $\rho(E_0)$ is a normal Dehn disk without triple points, and E is obtained from $\rho(E_0)$. $\square$

There may be occasions when one does not object to using discard in simplify-ing canonical Dehn disks. In this case, one does not need to put tubes near triple points or branch points. Putting in these tubes was necessitated by the use of Theorem XVI.6.B since then tubes were inserted to connect 2-spheres K_j to a disk E_0 and avoid discard.

THEOREM XVI.8.B. *Suppose D is a normal Dehn disk in a PL 3-manifold-without-boundary M^3. Then D can be changed to a nonsingular disk E by cut, paste, and discard by adjusting near branch points, and cutting along double lines and through triple points and resewing. No tubes are added near either branch points or triple points.*   □

CHAPTER XVII

THE LOOP THEOREM

XVII.1. Versions of the loop theorem. Suppose M^3 is a connected PL 3-manifold-with-boundary. One might try to simplify M^3 by cutting off a handle. This would involve finding a PL disk D such that Bd $D \subset$ Bd M^3, Int $D \subset$ Int M^3, and $M^3 - D$ are connected. Let P^3 be a pillbox obtained by thickening D. The *pillbox* P^3 is the image of $D \times [-1, 1]$ under a PL homeomorphism h into M^3 such that each $h(x \times 0) = x$, $h(\text{Bd } D \times [-1, 1]) \subset$ Bd M^3, and $h(\text{Int } D \times [-1, 1]) \subset$ Int M^3. It is sometimes called a 1-handle. Technically there are 0-handles and 2-handles but we use them less frequently, and if we use the term *handle* without qualification, we mean it to be like the handle of a bucket—a cylinder $D^2 \times [0, 1]$ attached at its ends. This is a 1-handle or a pillbox. If $M^3 - P^3$ is connected, its fundamental group has one less generator than that of M^3. The loop theorem is sometimes useful in determining whether or not one can remove a handle from a 3-manifold-with-boundary.

Recall that a *loop in X* is a map of a simple closed curve J into X. Also, the loop *f can be shrunk to a point in X* if there is a disk D, a homeomorphism h of J onto Bd D, and a map g of D into X such that $f = gh$. We may think of $f(J)$ as being gradually shrunk to a point in X as time t varies between 0 and 1 if we regard D as a circular planar disk with radius 1 and decide that at time t the point $f(x)$ has been moved to $g(y)$ where y is the point of D on the radius to $h(x)$ which is at a distance of t from $h(x)$. At time 1, $f(J)$ has been shrunk to g (center of D).

One version of the loop theorem can be stated simply as follows.

THEOREM XVII.1.A (LOOP THEOREM). *Suppose M^3 is a PL 3-manifold-with-boundry (not necessarily compact) such that there is a loop on Bd M^3 that can be shrunk to a point in M^3 but not in Bd M^3; then there is a nonsingular loop with these properties.* □

With the help of Dehn's lemma, Theorem XVII.1.A can be extended to the following result.

THEOREM XVII.1.B. *Under the hypothesis of Theorem XVII.1.A there is a PL nonsingular disk D in M^3 such that $D \cap$ Bd $M^3 =$ Bd D but Bd D cannot be shrunk to a point in Bd M^3.* □

Recall that if $x_0 \in X$, then an element of $\pi_1(X, x_0)$ is an equivalence class of loops in X with end points at x_0 where here it is convenient to define a loop with end points at x_0 as a map $g: [0, 1] \to X$ such that $g(0) = g(1) = x_0$. If $f: [0, 1] \to X$ is a loop in X with end points at x_0, the *conjugacy class of f* is the set of all elements $[g] \in \pi_1(X, x_0)$ such that for some loop $g \in [g]$,

$$g(t) = g(1 - t) \quad \text{for } 0 \leqslant t \leqslant 1/3, \text{ and}$$
$$g(t) = f(3t - 1) \quad \text{for } 1/3 \leqslant t \leqslant 2/3.$$

In a certain sense, g resembles f on a circular stem. Also, if $f(S^1)$ is a singular closed curve (or J a simple closed curve) in X we say that $[g]$ belongs to the conjugacy class of $f(S^1)$ (or J) if we think of S_1 as $[0, 1]$ with the ends identified and find that $[g]$ belongs to the conjugacy class of $f: S^1 \to X$ (or Identity: $J \to X$). We denote the conjugacy class of $f, J, f(S^1)$ by $C(f, X)$, $C(J, X)$, $C(f(S^1), X)$.

In the statement of Theorem XVII.1.A, the condition that there is a loop in Bd M^3 that can be shrunk to a point in M^3 but not in Bd M^3 may be abbreviated to say that there is a component B of Bd M^3 such that $\pi_1(B) \to \pi_1(M^3)$ has a nontrivial kernel.

While the version of the loop theorem given by Theorem XVII.1.A is useful and easy to state, it may actually be easier to prove the following stronger version that is more difficult to state.

THEOREM XVII.1.C. *Suppose M^3 is a 3-manifold-with-boundary (not necessarily compact) and B is a component of Bd M^3. If N is a normal subgroup of $\pi_1(B, x_0)$ and f is a loop in B such that*
   f can be shrunk to a point in M^3 but
   no element of N lies in the conjugacy class of f.
Then there is a nonsingular loop g in B such that

   g can be shrunk to a point in M^3 but
   no element of N lies in the conjugacy class of g. □

Theorem XVII.1.C can be stated alternatively as follows.

THEOREM XVII.1.D. *Suppose M^3 is a 3-manifold-with-boundary (not necessarily compact), B is a component of Bd M^3, and N is a normal subgroup of $\pi_1(B)$ such that Kernel$[\pi_1(B) \to \pi_1(M^3)] \not\subset N$. Then there is a simple closed curve J in B such that*

$$\text{Kernel}\left[\pi_1(B) \to \pi_1(M^3)\right] \supset C(J, B) \not\subset N.$$ □

The normal Dehn disk was permitted to have three kinds of singularities: double lines, triple points, and branch points. The singular disks one uses in the loop theorem have singularities on the boundary of the 3-manifold so we also

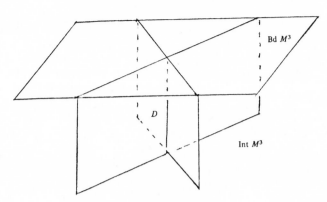

Figure XVII.1.D.

permit singularities of the sort shown in Figure XVII.1.D. This type of singularity is called the *end of a double line*.

Suppose M^3 is a PL 3-manifold-with-boundary and f is a PL map of Δ^2 into M^3. We call $f(\Delta^2) = D$ a *normal disk* if $D \cap$ Bd $M^3 =$ Bd D and the only singularities of D are double lines, triple points, branch points, and ends of double lines. In the next two sections we prove the following cut, paste, and discard version of the loop theorem. A small general position adjustment can change any singular PL disk D in M^3 with $D \cap$ Bd $M^3 =$ Bd D to a normal disk. Such an adjustment could be obtained by making Bd D normal and selecting f so that for each point $p \in$ Bd Δ, f is a homeomorphism on some open subset of Δ^2 containing p.

We shall prove the following version of the loop theorem and that will imply Theorems XVI.1.A, XVI.1.B, XVI.1.C, and XVI.1.D.

THEOREM XVII.1.E. *Suppose D is a normal singular disk in a PL 3-manifold-with-boundary M^3, B is a boundary component of Bd M^3, and N is a normal subgroup of $\pi_1(B)$ such that no element of N belongs to the conjugacy class of Bd D. Then D can be changed by cut, paste, and discard to a nonsingular disk E such that Bd $E \subset B$ and no element of N belongs to the conjugacy class of Bd E.*

EXAMPLES. One might hope to obtain the disk E of the conclusion of Theorem XVII.1.E by cut and paste without discard. However, this is not always possible. Let M^3 be a solid torus and D be a cone from a point of Int M^3 over a normal closed curve on Bd M^3 which goes around Bd M^3 twice meridianally and has only one singularity. If one used cut and paste without discard to remove singularities, one gets either a nonsingular disk E such that Bd E can be shrunk to a point in Bd M^3 or else one gets an annulus whose boundary components cannot be shrunk to points on Bd M^3.

Our second example is shown in Figure XVII.1.E where M^3 is a solid double torus (cube with two handles) and N is the minimal normal subgroup of $\pi_1($Bd $M^3)$ containing $C(J, $Bd $M^3)$, where J is the curve circling the right handle of Bd M^3 meridionally as shown in the left part of Figure XVII.1.E. The right part shows D schematically. When D is changed by cut, paste, and discard to E so

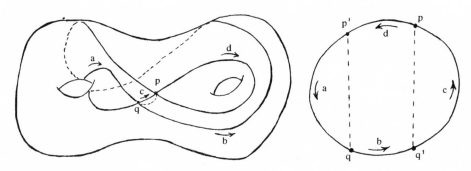

Figure XVII.1.E

that $C(\text{Bd } E, \text{Bd } M^3) \cap N = \varnothing$, a Moebius band is left over. It takes a bit of checking to see that the other sewing leads to an E so that $C(\text{Bd } E, \text{Bd } M^3) \subset N$.

XVII.2. Eliminating branch points. If M^3 is a PL 3-manifold-with-boundary and $f(\Delta^2) = D$ is a normal disk in M^3, then by adjusting D in a neighborhood of its branch points, one can use the methods of §XVI.4 to get a new normal disk D_1 in M^3 such that each branch point is of order 1.

Suppose p is a branch point of this simplified normal disk D_1. If one starts at p and begins splitting and resewing as suggested in the proof of Theorem XVI.4.B, one may be blocked by arriving at a second branch point q. In this case, D_1 can be adjusted in a neighborhood of q as suggested in the proof of Theorem XVI.4.B and eliminate q as a branch point.

If this splitting and resewing does not lead to another branch point, it leads to the end of a double line. The inverse of this short double line is a spanning arc of Δ^2 that divides Δ^2 into two disks. If one continues this splitting and resewing through this end of the double line, one can do this so as to obtain two normal disks E_1, E_2.

Suppose B is the component of Bd M^3 containing Bd D_1 and N is a normal subgroup of $\pi_1(B)$ such that $C(\text{Bd } D_1, B) \cap N = \varnothing$. If there are elements $g_1 \in C(\text{Bd } E_1, B)$, $g_2 \in C(\text{Bd } E_2, B)$ in N, then by adjusting g_2 (to put it on the same stem as g_1) one finds that there is an alement g_2', in $C(\text{Bd } E_2, B) \subset N$ such that $g_1 g_2' \in C(\text{Bd } D_1, B)$. The existence of such a $g_1 g_2'$ contradicts the requirement that $N \cap C(\text{Bd } D_1, B) = \varnothing$. Hence, we suppose with no loss of generality that $C(\text{Bd } E_1, B) \cap N = \varnothing$.

In a finite number of cut, paste, and discard steps, all branch points can be removed from our singular disk.

THEOREM XVII.2. *Suppose M^3, B, N, D are as in the statement of Theorem XVII.1.E. Then D can be changed to a normal disk E with cut, paste, and discard so that E has no branch points,* Bd $E \subset B$, *and $C(\text{Bd } E, B) \cap N = \varnothing$.* $\square$

XVII.3. Thickening canonical disks. A normal disk D in a 3-manifold-with-boundary M^3 is called *canonical* if it has no branch points. We learn from Theorem XVII.2 that in the proof of Theorem XVII.1.E we can restrict our attention to canonical disks.

Suppose $M^3(D)$ is the 3-manifold-with-boundary obtained by thickening $f(\Delta^2)$ $= D$ in M^3 as suggested in §XVI.6. A difference is that since Bd $D \subset$ Bd M^3, we suppose that the thickening of Bd D is a 2-manifold-with-boundary A in Bd M^3. Then A is a regular neighborhood in Bd $M^3(D)$ of Bd D and is a disk with handles (which are not known to be orientable) which has Bd D as a spine.

We shall now direct our attention to $M^3(D)$ rather than M^3 and to a normal subgroup $N(A)$ of $\pi_1(A)$ rather than N. Let $N(A)$ be the inverse of N under the injection $\pi_1(A) \to \pi_1(B)$. If $g^* \in C(\text{Bd } D, A)$, then the image of g^* under the injection is in $C(\text{Bd } D, B)$. Hence, $C(\text{Bd } D, A) \cap N(A) = \varnothing$. Likewise, when we change D to a canonical disk E so that $N(A) \cap C(\text{Bd } E, A) = \varnothing$, then $N \cap C(\text{Bd } E, B) = \varnothing$ also.

THEOREM XVII.3. *Suppose $M^3(D)$ has the property that A lies in a disk E_0 in* Bd $M^3(D)$ *and each connected PL 2-manifold-without-boundary in* Int $M^3(D)$ *separates $M^3(D)$. Then by cut, paste, and discard one can change D to a nonsingular disk E so that* Bd $E \subset A$ *and $N(A) \cap C(\text{Bd } E, A) = \varnothing$.*

PROOF. We suppose Bd $E_0 \subset$ Bd A and $E_1, E_2, \ldots, E_n$ are the closures of the components of $E_0 - A$. If for each $i = 1, 2, \ldots, n$, $N(A) \cap C(\text{Bd } E_i, A) \neq \varnothing$, then $N(A) \cap C(\text{Bd } D, A) \neq \varnothing$. Hence there is an E_i (say E_1) such that $N(A) \cap C(\text{Bd } E_1, A) = \varnothing$. If we were not concerned with cut and paste we could use E_1 for the E required by the theorem. Note that E_1 projects naturally into a singular disk E_1' in the point set D in $M^3(D)$. Since each connected orientable 2-manifold-without-boundary in Int $M^3(D)$ separates $M^3(D)$, it follows as it did in the proof of Theorem XVI.6.B that for no $x \in \Delta^2$ do both ends of the fiber through $f(x)$ lie in E_1. It follows from the techniques used in the proof of Theorem XVI.6.A that one can split E_1 (and hence D) along these singularities and resew, so as to obtain a nonsingular disk E such that Bd E is homotopic to Bd E_1 in A. Hence, one can use cut, paste, and discard to change D to a nonsingular disk E such that $N(A) \cap C(\text{Bd } E, A) = \varnothing$. $\square$

XVII.4. Proof of loop theorem. Let M^3 be a PL 3-manifold-with-boundary and $d(\Delta^2) = D$ a canonical disk in M^3 with boundary on Bd M^3. Our proof of the loop theorem is modeled after the proof of the Dehn's lemma given in §XVI.8 and is by induction on $n(d)$. We assume a familiarity with that proof and omit some details here. Recall that $n(d)$ is the number of 1-simplexes in a minimal curvilinear triangulation of $S(d)$ in Δ^2.

Assume $d_0(\Delta^2) = D_0$ is a canonical disk which is a counterexample to the loop theorem such that $n(d_0)$ is minimal. Let $M^3(D_0)$ be the result of thickening D_0, $A_0 = $ Bd $M^3 \cap M^3(D_0)$, and $N(A_0)$ be the inverse of N under the homomorphism $\pi_1(A_0) \to \pi_1(B)$. It follows from Theorem XVII.3 that either A_0 does not lie in a disk in Bd $M^3(D_0)$ or some PL 2-manifold-without-boundary in Int $M^3(D_0)$ fails to separate $M^3(D_0)$. In either case, it follows from Theorem XV.5.C that $M^3(D_0)$ has a connected 2-sheeted covering $\tilde{M}^3(D_0)$ with its associated projection map $\rho: \tilde{M}^3(D_0) \to M^3(D_0)$.

Since D_0 can be lifted, there are maps d_1, d_2: $\Delta^2 \to \tilde{M}^3(D_0)$ such that each $\rho d_i = d_0$ and $\rho^{-1}(D_0) = d_1(\Delta^2) \cup d^2(\Delta^2)$. Then $d_1(\Delta^2)$ is a canonical disk in $M^3(D_0)$ and $n(d_1) < n(d_0)$. Let $\tilde{A}$ be a regular neighborhood of $d_1(\text{Bd } \Delta^2)$ on Bd $\tilde{M}(D_0)$ and $\tilde{N}$ be the normal subgroup of $\pi_1(\tilde{A})$ which is the inverse of $N(A_0)$ under the projection homomorphism. Note that $\tilde{N} \cap C(d_1(\text{Bd } \Delta^2), \text{Int } \tilde{A}) = \varnothing$. Our induction on $n(d)$ implies that by using cut, paste, and discard one can change $d_0(\Delta^2)$ to a nonsingular disk $e_1(\Delta^2)$ so that $C(e_1(\text{Bd } \Delta^2), \text{Int } \tilde{A}) \cap \tilde{N} = \varnothing$. Then $\rho e_1(\Delta^2) = e_0(\Delta^2)$ is a canonical disk E_0 in $M^3(D_0)$ which can be obtained from D_0 by cut, paste, and discard. Also, $C(\text{Bd } E_0, A_0) \cap N(A_0) = \varnothing$.

Since ρ is a 2-sheeted covering, E_0 has no triple points. The techniques used in the proof of Theorem XVI.3.A enables one to remove all the simple closed curves from $S(E_0)$. Hence we suppose that $S(E_0)$ is the union of arcs, and $S(e_0)$ is the union of spanning arcs p_1q_1, $p_1'q_1'$, $p_2q_2,\ldots,p_n'q_n'$ of Δ^2 and these arcs appear in pairs in the sense that $e_0(p_iq_i) = e_0(p_i'q_i')$.

Let us consider how we can eliminte $e_0(p_1q_1) = e_0(p_1'q_1')$ from $S(E_1)$. There are two cases according as how $e_0(p_1)$, $e_0(q_1)$, $e_0(p_1')$, $e_0(q_1')$ are ordered on Bd E_1. The top part of Figure XVII.4.a gives a schematic view of E_0 under one ordering while the bottom part gives a schematic view of the other. In either case, let P_p be a path from x_0 to $e_0(p_1) = e_0(p_1') = p$ in A_0 and P_q be a path from x_0 to

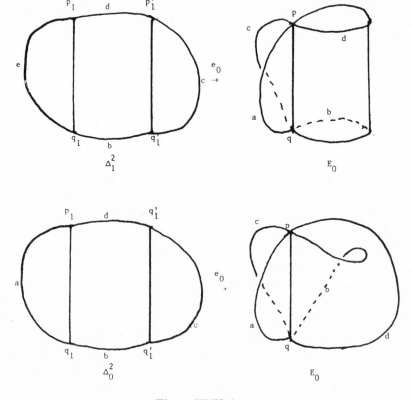

Figure XVII.4.a

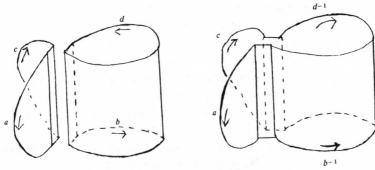

Figure XVII.4.b

$e_0(q_1) = e_0(q_1') = q$ in A_0. Let a, b, c, d be elements of $\pi_1(A_0, x_0)$ containing the loops that start at x_0, go out P_p or P_q, then along Bd E_0 as indicated in the figure by a, b, c, d, and then return to x_0 by way of P_q or P_p. The conjugacy class of Bd E_0 contains $abcd$. Hence $abcd \notin N(A_0)$.

Figure XVII.4.b gives a view of how the E_0 as described by the top part of Figure XVII.4.a is changed to get a nonsingular disk. We still do not show $e_0(p_iq_i)$ for $i > 1$. There are two ways to slice and resew E_0 along $e_0(p_1q_1) = e_0(p_1'q_1') = pq$ as shown in Figure XVII.4.b. In the left part of Figure XVII.4.b, splitting and resewing is done so that an annulus is discarded and a smaller canonical disk remans while in the right part there is no discard. The conjugacy class of the boundary of the first disk contains ac while that of the second contains $ab^{-1}cd^{-1}$. If we show that one of ac, $ab^{-1}cd^{-1}$ is not in $N(A_0)$, then we have shown how to simplify E_0 of the top part of Figure XVII.4.a.

If $ac \in N(A_0)$, we can essentially interchange a with c^{-1} and a^{-1} with c in determining whether or not an element is in $N(A_0)$. For example, if each of ac, $ab^{-1}cd^{-1}$ is in $N(A_0)$, then we find successively that each of $dc^{-1}ba^{-1}$, $(dc^{-1}ba^{-1})ac$, $d^{-1}(dc^{-1}bc)d$, $ac(c^{-1}bcd)$, $abcd$ in $N(A_0)$. Having $abcd$ in $N(A_0)$ contradicts the condition that $C(\text{Bd } E_0, A_0) \cap N(A_0) = \varnothing$. Hence, we can simplify the E_0 of the top part of Figure XVII.4.b by cut, paste, and possibly discard.

Had the ordering of $e_0(p_1)$, $e_0(q_1)$, $e_0(p_1')$, $e_0(q_1')$ been as shown in the lower part of Figure XVII.4.a, then the splitting and resewing along $e_0(p_1q_1) = e_0(p_1'q_1') = pq$ could have been done in two ways. The conjugacy classes of the boundaries of the resulting disks would have contained ac^{-1}, $adbc$, respectively. An exercise in group theory shows that if each of these belongs to $N(A_0)$, then so does $abcd$. (Recall that if $ac^{-1} \in N$, one can essentially interchange a with c in determining whether or not an element belongs to N.) The lower part of Figure XVII.4.a is related to the example described with Figure XVII.1.E.

In any case, we find that we can get a simpler disk by splitting and resewing along pq. In a finite number of such operations one can change E_0 to a nonsingular disk E such that $C(\text{Bd } E, A_0) \cap N(A_0) = \varnothing$. The existence of this E establishes Theorem XVII.1.E, since it shows that there is no D_0 as assumed. We have shown that D_0 can be changed by cut, paste, and discard to a nonsingular disk E such that $C(\text{Bd } E, B) \cap N = \varnothing$. $\square$

Now that we have finished the proof of Theorem XVII.1.E, it is natural to wonder why it was convenient to require that $N \cap C(\mathrm{Bd}\ D, B) = \varnothing$ rather than that Bd D did not shrink to a point in Bd M^3. In our proof of Theorem XVII.1.E we shifted our attention from M^3 to $M^3(D_0)$ to $\tilde{M}^3$ and shifted it from B to A to A_0. Suppose we wanted to avoid normal subgroups and just prove Theorem XVII.1.A or XVII.1.B. In building E_0 it is not enough that Bd E_0 does not shrink in A—it must not shrink in B either. We do not want merely that $C(\mathrm{Bd}\ E_0, A)$ not lie in the identity element of $\pi_1(A)$—we want it not to lie in a certain normal subgroup of $\pi_1(A)$—namely the one containing equivalence classes of loops in $\pi_1(A)$ that do not shrink in B. We were better able to keep track of whether or not certain replaced normal subgroups intersected a replaced conjugacy class of Bd D than to look back at the original Bd M^3.

XVII.5. History. Kneser gives a specialized form of the loop theorem in [K_4]. In the statement of a lemma (Hilfssatz) is contained both the loop theorem and a consequence of Dehn's lemma, the proof of which illuminated the celebrated hiatus in Dehn's proof. Whitehead gave the same specialized form of the loop theorem in a knot-theoretic setting in [W_5]. This proof was more algebraic in flavor, and it tied down what Whitehead called an "unconvincing" paragraph in Kneser's proof.

In 1957 Papakyriakopolous [P_2] proved a more general form of the loop theorem. In a later expository paper, Papakyriakopolous [P_3] gave an enlightening account of the parallel functions of Dehn's lemma, the loop theorem, and the sphere theorem in the classification problem for manifolds.

The form of the theorem appearing in Theorem XVII.1.D was presented in 1959 by J. Stallings [S_6] and extended the loop theorem to nonorientable manifolds. This proof uses the method of double-sheeted coverings which was introduced earlier by Shapiro and Whitehead in [SW_1] to give a proof of Dehn's lemma. Henderson further refines the loop theorem in [H_8] by techniques similar to those used in this chapter. Other treatments are given by Moise [M_{11}] and Hempel [H_7].

CHAPTER XVIII

RELATED RESULTS

In the final chapter of this book we list some results related to material considered in earlier chapters. It would require much space to give complete coverage. Many proofs would be repetitious of material already well treated elsewhere. It is hoped that many who have gotten this far can supply some of their own details. We shall limit ourselves in this chapter to listing results, discussing overall strategy of proofs, and giving references.

XVII.1. Approximating 2-complexes. Theorem XIII.A about approximating 2-spheres with PL spheres can be adjusted to give a result about approximating disks with PL disks. A set Y in a triangulated M is *locally polyhedral* at y if $y \in Y \subset M$, and there is a neighborhood N of y in M such that $\bar{N} \cap Y$ is a finite polyhedron in M.

THEOREM XVIII.1.A. *Suppose D is a topological disk in E^3 and f is a positive continuous function defined on* Int *D. Then there is a homeomorphism h of D onto a topological disk D' in E^3 such that* Bd $D =$ Bd D', $d(x, h(x)) \leq f(x)$ *for each point x of D,* Int *D' is locally polyhedral, and D' is locally polyhedral at each point of* Bd *D' at which* Bd *D is locally polyhedral.*

If K is a closed subset of D at which D is locally polyhedral, h may be chosen so as to be the identity on K.

OUTLINE OF PROOF. The proof can be a modification of that of Theorem XIII.C. Details are found in the proof of $[\mathbf{B_{11}}$, Theorem 1]. It is also found in $[\mathbf{B_9}$, Theorems 7 and 8]. $\square$

Harrold $[\mathbf{H_3}]$ made one of the earliest applications of the approximation theorem when he used it to strengthen a characterization of tame arcs in R^3 that had been given earlier with unneeded restrictions.

Theorem XVIII.1.A can be extended to apply to 2-complexes as follows.

THEOREM XVIII.1.B (APPROXIMATION THEOREM FOR 2-COMPLEXES). *Suppose M is a triangulated 3-manifold-with-boundary, P is a closed set in M which is the image of a 2-complex C under a homeomorphism g, and f is a positive continuous function defined on P. Then there is a homeomorphism h of P onto a polyhedron P' in M such that $d(x, h(x)) < f(x)$.*

If K is a closed set on which P is locally polyhedral, h may be chosen so as to be the identity on K.

For each triangulation of C, h may be chosen so that $hg(\Delta) \subset$ Bd M for each simplex Δ in this triangulation for which $g(\Delta) \subset$ Bd M.

OUTLINE OF PROOF. Details of the extension are as found in the treatment of [B_{11}, Theorem 5]. Theorem XVIII.1.A is used to make the image of the 2-complex locally PL on the interiors of its 2-faces. Then scissors-and-paste techiques are used to adjust it further so that it becomes locally PL on the interiors of 1-faces. Finally it is adjusted to make it locally PL at the vertices. □

There is a version of Theorem XVIII.1.B that only requires that f be a nonnegative continuous function. In this version one only concludes that P' is locally polyhedral at $h(x)$ if $f(x) > 0$. This version can be obtained by removing from P and M the portion of P on which $f(x) = 0$ and applying Theorem XVIII.1.B to the remainder.

Instead of making P' locally polyhedral, one might construct an h such that hg is locally PL. A map ϕ of a complex C is *locally* PL at a point $p \in C$ if for some triangulation T of C, ϕ is linear on each simplex of T that contains p.

XVIII.2. Approximating homeomorphisms on 3-manifolds. Moise [M_8] was the first to show that if h is a homeomorphism of a 3-simplex Δ^3 into a PL 3-manifold-with-boundary M^3 and $\varepsilon > 0$, then there is a PL homeomorphism $g: \Delta^3 \to M^3$ such that $d(h, g) < \varepsilon$. Theorem XVIII.2.A is a modification of that result.

THEOREM XVIII.2.A. *Suppose M_1, M_2 are triangulated 3-manifolds-with-boundaries, f is a continuous positive function defined on an open subset U of M_1, and h is a homeomorphism of U onto M_2. Then there is a locally PL homeomorphism h' of U onto $h(U)$, such that $d(h(x), h'(x)) < f(x)$ for each point x of U.*

If K is a closed subset of U on which h is already locally PL, the homeomorphism h' may be chosen so that $h = h'$ on K.

OUTLINE OF PROOF. Details of proof are found in [B_{11}, Theorem 9]. The part of U in M_1 is subdivided to be very fine and Theorem XVIII.1.B is used to approximate the homeomorphism h of the 2-skeleton of the triangulation by a PL homeomorphism h'. Then the PL Schoenflies theorem (Theorem XIV.1) is used to define h' on the 3-simplexes of the triangulation. As an afterthought we note that we should have made elements of the subdivisions so small and f so small that for each 3-simplex Δ^3 of the subdivision, $h'(\text{Bd } \Delta^3)$ lies in a PL 3-cell in M^3. Also, since M^3 might have boundary we could suppose that no 2-simplex of the subdivision separates a component of U. (This keeps two parts of U from getting on the same side of a 2-simplex.) To make $h'(U) = h(U)$ we could make $h = h'$ on the 0-skeleton of the subdivision. □

Many uses have been made of the fact that any homeomorphism of one PL 3-manifold into another can be approximated by a PL homeomorphism. Samples are included in Moise [M_9, M_{11}], Sanderson [S_1] and Bing [B_7].

The following theorem is a corollary of Theorem XVIII.2.A.

THEOREM XVIII.2.B. *If T_1, T_2 are triangulations (perhaps not compatible) of the same 3-manifold-with-boundary M, there are subdivisions T_1', T_2' of T_1, T_2 and a linear homeomorphism of M under T_1' onto M under T_2'. We can pick the homeomorphism near the identity.* □

The following theorem also follows from Theorem XVIII.2.A.

THEOREM XVIII.2.C. *Each orientation preserving homeomorphism of R^3 onto itself is stable.* □

It would be interesting to know if Theorems XVIII.2.A, XVIII.2.B, or XVIII.2.C generalize to higher dimension.

XVIII.3. Triangulating 3-manifolds. If M^3 is a 3-manifold-with-boundary, it is the union of a locally finite collection of open sets $U_1, U_2, \ldots$, each of which can be triangulated. To show that M^3 can be triangulated requires a bit more than showing that $U_1 \cup U_2$ can be triangulated, but Theorem XVIII.3.A is a step in the right direction.

THEOREM XVIII.3.A. *Suppose U_1, U_2 are open subsets of a 3-manifold-with-boundary M and d_1, d_2 are metrics for U_1, U_2, respectively, that support triangulations T_1, T_2, respectively, of U_1, U_2. Then there is a metric d_0 for $U_1 \cup U_2$ that supports a triangulation T_0 of $U_1 \cup U_2$ such that each simplex of T_1 in $U_1 - \overline{U_2}$ and each simplex of T_2 in $U_2 - \overline{U_1}$ is in T_0.*

OUTLINE OF PROOF. Let $f(x)$ be a continuous function defined on M that is positive on $U_1 \cap U_2$ and zero elsewhere. Let d be a metric for M and T_i', $i = 1, 2$, be triangulations of $U_1 \cap U_2$ such that if x is a point of a simplex Δ of T_i, diameter of Δ under d is less than $f(x)$ and each element Δ of T_i' lies linearly in a simplex of T_i. Let h be the PL homeomorphism promised by Theorem XVIII.2.A of $U_1 \cap U_2$ under T_1' onto $U_1 \cap U_2$ under T_2' and such that $d(x, h(x)) < f(x)$ for each x. We extend h to be the identity on $(U_1 \cup U_2) - (U_1 \cap U_2)$.

We now seek a set that separates $U_1 - \overline{U_2}$ from $U_2 - \overline{U_1}$ and triangulations of the sides of the separator. To get the separator we let X be the union of the 3-simplexes of T_1 that intersect $U_1 - U_2$, B be the union of the 2-simplexes of T_1 in X that are faces of a 3-simplex of T_1 not in X and $h(B)$ be the required separator. Note that both B and $h(B)$ lie in $U_1 \cap U_2$. Since h is PL on B, there is a triangulation $T(B)$ of B such that $T(B)$ refines T_1 and h takes each simplex of $T(B)$ linearly into a simplex of T_2. Following the technique of Theorem I.2.A we get a triangulation T_1' of X such that each 3-simplex of T_1 in $U_1 - U_2$ and each element of $T(B)$ is an element of T_1''. The images under h of simplexes of T_1'' are the simplexes of T_3 on the U_1 side of $T(B)$.

The closure (in $U_1 \cup U_2$) of the other side of the separator $h(B)$ is triangulated so that the elements of T_2 in $U_2 - U_1$ are elements of T_3 as are the images under h of elements of $h(B)$.

The above proof is reminiscent of that of Theorem II.5. For a more detailed treatment see Bing [B_{11}] or Moise [M_{10}]. Alternative treatments are found in Shalen [S_2] and Hamilton [H_2].

THEOREM XVIII.3.B. *Each 3-manifold-with-boundary can be triangulated.*

PROOF. Suppose M^3 is a 3-manifold-with-boundary. Since each point of M^3 lies in a neighborhood O which is topologically equivalent to either R^3 or the points of R^3 on or above the xy-plane, there is a sequence of open sets $O_1, O_2, \ldots,$ covering M such that each O_i can be triangulated.

Let $U_1, U_2, \ldots,$ be an increasing sequence of open sets covering M^3 such that $\overline{U}_i$ is a compact subset of U_{i+1}. Since $\overline{U}_i$ can be covered by a finite number of the O_i's, it follows from Theorem XVII.3.A that U_i lies in a triangulated open subset of M^3 and can therefore be triangulated. Since each $U_{i+2} - \overline{U}_i$ can be triangulated, so can each of the open sets $U_2 \cup (U_4 - \overline{U}_2) + (U_6 - \overline{U}_4) + \cdots$ and $(U_3 - \overline{U}_1) + (U_5 - \overline{U}_3) + \cdots$. Again it follows from Theorem XVIII.3.B that their union can be triangulated. □

Theorem XVIII.3.B has the following extension.

THEOREM XVIII.3.C. *If M is a 3-manifold-with-boundary and $T(\text{Bd } M)$ is a triangulation of Bd M, then there is a triangulation $T(M)$ of M compatible with $T(\text{Bd } M)$ on Bd M.*

OUTLINE OF PROOF. One method of proof is to use Theorem XVIII.3.B to get a triangulation T of M which need not be compatible with $T(\text{Bd } D)$ on Bd M. Then note that there is an open set in M obtained as a collar of Bd M. Modify T in a feathered fashion on this cartesian product neighborhood so as to get a required triangulation of M.

An alternative proof is provided by [B_7, Theorem 5]. □

The following is a variation of Theorem XVIII.3.A.

THEOREM XVIII.3.D. *Suppose M is a 3-manifold-with-boundary, U is an open subset of M, X is a closed set in X, $T(M)$ is a triangulation of M, and $T(U)$ is a triangulation of U (not necessarily compatible with $T(M)$). Then there is a triangulation T of M such that each simplex of $T(U)$ intersecting X is an element of T as is each simplex of $T(M)$ missing U.* □

XVIII.4. Locally tame sets are tame. Recall that a closed subset K of a PL 3-manifold-with-boundary M is *tame* if there is a homeomorphism $h: M \to M$ such that $h(k)$ is a polyhedron. It is *locally tame* in M if for each point $p \in K$ there is a neighborhood N_p of P in M and a homeomorphism $h: M \to M$ such that $h_p(\overline{N}_p \cap K)$ is a polyhedron. Bing [B_7] and Moise [M_{11}] showed closed subsets of a PL 3-manifold were tame if they were locally tame. As a step toward the proof of that result we call attention to [B_7, Theorem 8] which we quote as follows.

THEOREM XVIII.4. A. *If K is a locally tame closed subset of a 3-manifold-with-boundary M, then M has a triangulation under which K is a polyhedron.*

OUTLINE OF PROOF. A proof is given [B_7, pp. 154–156], but we outline one below. First we note from Theorem XVIII.3.B that M has a triangulation T_1. Theorem XVIII.3.D is used to modify T_1 to get a triangulation T_2 of M such that K is locally polyhedral under T_2 at each point of K that is destined by the local structure to be a vertex of any triangulation of M making K polyhedral. Then polyhedral neighborhoods of these special vertex points are removed so that the remainder is a 3-manifold-with-boundary. Then this remaining part is split along the 2-dimensional part of the remaining part of K. The boundary of this reduced and split 3-manifold-with-boundary is triangulated to get a fit when resewing and reattachment is done. Theorem XVIII.3.B is used to extend this triangulation of the boundary to the reduced and split 3-manifold-with-boundary. After the resewing and reattachment there results a triangulation T_3 of M such that K is locally polyhedral except possibly along some arcs. It is shown in Lemma 6 and Theorem 7 [B_7, pp. 153–154] how to pull locally tame arcs onto PL arcs.   □

The following version of Theorem XVIII.4.A is given by [B_7, Theorem 9].

THEOREM XVIII.4.B. *Suppose M is a triangulated 3-manifold-with-boundary, K is a closed set in M, and X is a closed subset of K such that K is locally polyhedral at each point of X and locally tame at each point of $K - X$. Then for each nonnegative continuous function $\varepsilon(x)$ on M such that $\varepsilon(x) > 0$ for each $x \in K - X$ there is a homeomorphism $h: M \to M$ such that $h(K)$ is a polyhedron and $d(x, h(x)) \leqslant \varepsilon(x)$.*   □

XVIII.5. Tameness from the side.

Suppose S is a tame 2-sphere in R^3, Int S is the bounded component of $R^3 - S$, and $\varepsilon > 0$. Then there is a homeomorphism h of S into Int S such that $d(h, \text{Id}) < \varepsilon$. But what if we are not given that S is tame but rather that for each $\varepsilon > 0$ there is an h. Can be conclude that S bounds a 3-ball? The answer is yes as given by the following result.

THEOREM XVIII.5.A. *Suppose in R^3 that S is a 2-sphere such that for each $\varepsilon > 0$ there is a homeomorphism h of S into Int S such that $d(h, \text{Id}) < \varepsilon$ on S. Then S bounds a 3-ball in R^3.*

OUTLINE OF PROOF. The proof is given in [B_{12}, Theorem 2.1]. We use the approximation theorem.

It is shown in [B_{12}, Theorem 3.1] that there is of sequence of mutually disjoint polyhedral 2-spheres $S_1, S_2, \ldots$, in Int S and homeomorphisms $h_1, h_2, \ldots$, of S onto the S_i's so that $d(h_i, \text{Id}) < 1/2^i$. It is shown further that the S_i's and h_i's could be chosen so that each adjacent pair (S_i, S_{i+1}) bounds a thin hollow ball so that there is a homeomorphism f of $S \times (0, 1/2]$ into Int S so that $f(x \times 1/2^i) = h_i(x)$ and diameter $f(x \times [1/2^i, 1/2^{i+1}]) < 1/2^i$. The homeomorphism h extends to $S \times [0, 1/2]$, so that the union of S, S_1, and the part of R^3 between them is a hollow ball. Theorem XIV.1 shows that S_1 bounds a 3-ball so S bounds a 3-ball with a collar.   □

EXAMPLE. One should not conclude that the S of Theorem XVIII.5.A is tame because it might be wild from the outside as described in §IV.3.

What happens in Theorem XVIII.5.A if instead of supposing that h_i is a homeomorphism we suppose it is a map. The answer is not at hand and leaves open the free surface problem which may be stated as follows.

Free surface problem. Suppose S is a 2-sphere in R^3 and for each $\varepsilon > 0$ there is a map g_ε of S into Int S such that $d(\text{Id}, g_\varepsilon) < \varepsilon$. Does this imply that S bounds a 3-ball?

This question may be related to the search for a better proof of Dehn's lemma where we want to do the cut-and-paste without getting long feelers.

Question. Suppose S is a round 2-sphere in R^3, and $\varepsilon > 0$. Is there a $\delta > 0$ such that if f is a map of S into R^3 with $d(\text{Id}, f) < \delta$, then each neighborhood of $f(S)$ contains the homeomorphic image $h(S)$ of S such that $d(h, \text{Id}) < \varepsilon$?

While we do have difficulty deciding if $f(S)$ is tame from Int(S) if it can be approximated by singular S's, we know that it is tame if it can be homotoped to the inside.

THEOREM XVIII.5.B. *A 2-sphere S in R^3 bounds a 3-ball in R^3 if there is a map of $S \times [0, 1]$ into $S \cup \text{Int}(S)$ such that for each $x \in S$, $h(x \times [0, 1]) \cap S = \{x\} = h(x \times 0)$.*

The proof is given in Hempel [**H**$_5$, Theorem 1]. □

If S is a 2-sphere in R^3, the side approximation theorem XIII.B says that there is a homeomorphism of S into a polyhedral 2-sphere S' such that S' almost lies in Int S. If it should happen that Int S is 1-ULC, disks of S' not in Int S could be replaced by singular disks so as to get a singular approximation. Scissors-and-paste can be used in this special situation so as to use Dehn's lemma and not develop feelers so we get the following result.

THEOREM XVIII.5.C. *Suppose S is a 2-sphere in R^3 such that Int S is 1-ULC. Then S bounds a 3-ball in R^3.*

The proof follows from [**B**$_{14}$, Theorem 1] and [**B**$_{12}$, Theorem 2.1]. □

A set X is *locally simply-connected* (1-LC) *at a point* $p \in \overline{X}$ if for each neighborhood U of p there is a neighborhood V of p such that each map of Bd Δ^2 into $V \cap X$ can be extended to map Δ^2 into $U \cap X$.

[**B**$_{14}$, Theorem 7] gives the following modification of Theorem XVIII.5.C.

THEOREM XVIII.5.D. *A 2-manifold M_2 in a triangulated 3-manifold M_3 is tame if and only if $M_3 - M_2$ is locally simply-connected at each point of M_2.* □

Suppose that M^2 is a 2-manifold in a PL 3-manifold M^3 and U is a component of $M^3 - M^2$. We say that M^2 is *tame from the U side* if U is 1-LC at each point of M^2.

THEOREM XVIII.5.E. *A 2-manifold in a PL 3-manifold is tame if it is tame from both sides.* □

By blistering or splitting a disk to change it to a 2-sphere one obtains the following application of Theorem XVIII.5.D.

THEOREM XVIII.5.E. *A disk in a 3-manifold is tame if its complement is* 1-*ULC.*

Details are found in [**B**$_{14}$]. □

The notion of taming sets has been very useful in studying the tameness of surfaces. A closed set X in R^3 is a *taming set* if it lies on some 2-sphere in R^3, and for each 2-sphere K^2 in R^3 containing X is tame if K^2 is locally tame at each point of $K^2 - X$. The following contain important results about taming sets: Bing [**B**$_{21}$, **B**$_{22}$], Burgess [**B**$_{31}$], Burgess and Cannon [**BC**$_1$], Cannon [**C**$_2$, **C**$_3$], Doyle and Hocking [**DH**], Gillman [**G**$_1$], Griffith [**G**$_4$], Loveland [**L**$_3$] and Moise [**M**$_{10}$]. Other papers dealing with the tameness of surfaces include Burgess [**B**$_{32}$], Burgess and Loveland [**BL**], Cannon [**B**$_4$], Eaton [**E**$_1$, **E**$_3$], Harrold [**H**$_4$] and Wilder [**W**$_7$].

XVIII.6. Reembedding crumpled cubes. Suppose C is a crumpled cube in R^3 and h_1 is a homeomorphism of C into R^3 such that h_1 is close to the identity and $h_1(\text{Bd } C) = S_1$ almost lies in $R^3 - C$. Using the technique of pulling back feelers as given in Chapter X one find that there is a homeomorphism g_1 of C into Int S_1 such that g_1 is close to the identity. Repeated use of this technique leads to the following result proved as Theorem X.3.B.

THEOREM XVIII.6.A. *Suppose C is a crumpled cube in R^3. Then for each $\varepsilon > 0$ there is a homeomorphism h of C into R^3 such that $R^3 - h(C)$ is* 1-*ULC, $d(h, \text{Id}) < \varepsilon$, and h is the identity outside the ε-neighborhood of* Bd C. □

The above is Theorem X.3.B and a proof is given in Chapter X. The proof was first obtained by Hosay [**H**$_{10}$] and Lininger [**L**$_1$]. Also see Daverman [**D**$_4$]. Hosay's announcement as given in [**H**$_{10}$] was not followed by a published paper, but it was as explained in [**B**$_{27}$].

The following interesting application of Theorem XVII.6.A is a restatement of Theorem X.4.A.

THEOREM XVIII.6.B. *If C is a crumpled cube and B is a 3-ball, the union of B and C sewed together with a homeomorphism between their boundaries is a 3-sphere.* □

Another version of Theorem XVIII.6.A is the following.

THEOREM XVIII.6.C. *Suppose X is a closed subset in a 3-manifold M^3 such that* Bd X *in M^3 is a 2-manifold. Then for each $\varepsilon > 0$ there is a homeomorphism h of X into M^3 such that $d(h, \text{Id}) < \varepsilon$, h is the identity outside the ε-neighborhood of* Bd X, *and $M^3 - h(X)$ is locally simply-connected at each point of* Bd(X). □

A *Sierpiński curve X* on a 2-sphere S is what remains in S after removing from S the interiors of a null sequence of mutually disjoint disks whose union is dense in S. A subset of X is said to be in the *inaccessible part* of X if it does not intersect the boundary of any of the removed open disks. It is known that a 2-sphere in a

3-manifold is tame if it is tame modulo a tame Sierpeński curve [$\mathbf{B_{21}}$]. Doyle [$\mathbf{D_3}$] and Doyle with Hocking [$\mathbf{DH}$] had shown that it is tame if it is tame modulo a tame arc. Also, if X_1, X_2 are two Sierpeński curves on 2-spheres S_1, S_2, any homeomorphism of X_1 onto X_2 can be extended to take S_1 onto S_2. We call a Sierpiński curve *tame* in a 3-manifold if it lies on a tame 2-sphere in the 3-manifold.

We now consider another modification of Theorem XVIII.6.A.

THEOREM XVIII.6.D. *For each* $\varepsilon > 0$ *the homeomorphism h of Theorem* XVIII.6.A *could have been selected so that there is a tame Sierpiński curve X on* Bd *C such that each component of* $S - X$ *is of diameter less than* ε *and h is fixed on X.*

OUTLINE OF PROOF. Note that in the proof of Theorem XVIII.6.A we could have picked h_1 so that h_1 is fixed on Bd C except on a finite number of small mutually disjoint disks in Bd C. By controlling the sizes of disks in later steps we could have gotten X in Bd C so that h is fixed on X. □

The following refers not to the reembedding of a crumpled cube but rather to the reembedding of a neighborhood of a 2-sphere.

THEOREM XVIII.6.E. *Each 2-sphere topologically embedded in an arbitrary 3-manifold has a neighborhood that can be embedded in* R^3.

The above result is proved by McMillan [$\mathbf{M_6}$].

XVIII.7. Tame sets in wild surfaces. Suppose our goal in Theorem XVIII.6.A was not to reembed C but rather to reembed Bd C. Instead of the h_1, used in the proof of Theorem XVIII.6.A, the h_1 used here is fixed on Bd C except for some small holes; simple closed curves in $R^3 - C$ and far from C could be shrunk in $R^3 - h_1(C)$. If we now turn our attention to the bounded component of $R^3 - h_1(C)$ we find that there is a homeomorphism $g_1 : h_1(C) \to R^3$ so that g_1 is the identity except on very small holes; small simple closed curves in $R^3 - h_1(C)$ (except for those very near $h_1(C)$) could be shrunk to points on small sets in $R^3 - g_1 h_1(C)$. By alternating between unbounded and bounded components of 2-spheres one gets homeomorphisms $h_1, g_1, h_2, \ldots, g_n, \ldots$, so that $\lim g_i h_i \cdots g_1 h_1$ is a homeomorphism on Bd C the complement of whose image is 1-ULC. This procedure leads to the following.

THEOREM XVIII.7.A. *Suppose S is a 2-sphere in* R^3 *and* $\varepsilon > 0$. *Then there is a tame Sierpiński curve X in S such that each component of* $S - X$ *is of diameter less than* ε.

The proof is as suggested in the paragraph preceding the statement of the theorem. Details of proof are found in [$\mathbf{B_{17}}$]. □

The following is an immediate consequence of Theorem XVIII.7.A.

THEOREM XVIII.7.B. *Each surface in a 3-manifold contains a tame arc.*

The proof is found in [$\mathbf{B_{17}}$]. □

The Sierpiński curve X can be used to obtain the following two results.

THEOREM XVIII.7.C. *Each topological disk in a 3-manifold can be pierced by a tame arc.*

Details are found in [**B$_{18}$**]. □

THEOREM XVIII.7.C. *Suppose S is a 2-sphere and A is a tame arc in R^3. Then for each $\varepsilon > 0$ there is an isotopy $h_t : R^3 \to R^3$ $(0 \leqslant t \leqslant 1)$ such that $h_0 = \text{Id}$, A intersects $h_1(S)$ in only a finite number of points and pierces it at each of these points, $d(h_t, \text{Id}) < \varepsilon$ and $h_t = \text{Id}$ outside an ε-neighborhood of $S \cap A$.*

Details are included in the proof of [**B$_{23}$**, Theorem 2]. Note that this is a more powerful result than Theorem XI.2, but it is built on the basis of the side approximation theorem XIII.B while the proof of Theorem XI.2 depended on more elementary concepts. □

XVIII.8. Characterizations. What are necessary and sufficient conditions that a connected, locally connected, nondegenerate compact metric space M be a 3-sphere? We know that M is a 1-sphere if it is separated by each pair of its points. It is a 2-sphere if it is locally connected, separated by each simple closed curve in it, but separated by no pair of its point. This is the Kline sphere characterization which was treated in [**B$_2$, B$_3$**]. One might have expected that the result might have been extended to say that M is a 3-sphere if it is separated by each 2-sphere in it but not by any set in it topologically equivalent to a proper subset of a 2-sphere. However, homology spheres show that this expectation does not hold even in the case of every 3-manifold.

We do have some characterizations of 3-spheres by Bing [**B$_4$, B$_{19}$**] and Bing with Martin [**BM**], but they are not very satisfying. Some involve certain types of partitionings and some deal with shellings of cellular partitionings. However, none have the property of being a sharp, easily understood, readily applicable set of conditions.

We do have the following two near characterizations.

THEOREM XVIII.8.A. *A compact connected 3-manifold M is a 3-sphere if each simple closed curve in M lies in a 3-cell in M.*

The result is proved in [**B$_{10}$**]. Two things that prevent Theorem XVIII.8.A from being a desired characterization is that the supposition that M is a 3-manifold is strong, and even so, there is the hope that we can replace the condition that each simple closed curve in M lies in a 3-cell with the simpler condition that each such simple closed curve can be shrunk to a point in M. □

Perhaps we are hoping for a solution to the following question.

Question. If P is a polygon in a simply-connected PL 3-manifold M, does P lie on a collapsible 2-complex in M?

An affirmative answer to the question would prove the well-known Poincaré conjecture.

Theorem XVIII.8.A designates a certain 3-manifold as being a 3-sphere though it is not regarded as a useful characterization. Another result is the following.

THEOREM XVIII.8.B. *Suppose* A_1, A_2 *are two models of the solid Alexander horned sphere and h is a homeomorphism of* Bd A_1 *onto* Bd A_2. *Then if h is used to sew* A_1 *to* A_2 *along* Bd A_1 *and* Bd A_2, *the resulting union is a 3-sphere; that is,*

$$A_1 \cup_h A_2 = S^3.$$

The proof is given in [**B₅**].   □

Theorem XVIII.8.B can be used to get the following interesting results about periodic maps.

THEOREM XVIII.8.C. *There is a periodic homeomorphism of period two of a 3-sphere onto itself whose fixed point set is a wild 2-sphere.*   □

Theorem XVIII.8.C broke the ice in getting exotic periodic homeomorphisms of S^3. Results by Alford [**A₅**], Bing [**B₂₀**], Montgomery with Samelson [**MS**], and Montgomery with Zippen [**MZ**] followed.

The conjecture that any compact connected simply-connected 3-manifold is topologically a 3-sphere is called the Poincaré conjecture since Poincaré conjectured something somewhat like this. However, his original work [**P₄**] did not speak of 3-manifolds but rather of two compact combinatorial n-manifolds. He claimed they were topologically equivalent if their corresponding homology groups were equivalent. Poincaré gave a counterexample in [**P₅**] of his faulty claim. Whitehead gave a counterexample in [**W₄**] to a faulty claim in [**W₃**]. There have been many faulty proofs of the Poincaré conjecture (see [**K₅**], for example) but most of them have not reached print. The conjecture has been discussed also in §§IV.2, VI.10, and XIV.6.

It is known that a compact connected 3-manifold has a *Heegaard splitting*—the manifold is the union of two *handlebodies* (cubes with handles) sewed together with a homeomorphism between the boundaries of the handlebodies. Efforts have been made to solve the Poincaré conjecture by simplifying Heegaard splittings. For example, see [**H₁**] by Wolfgang Haken and [**W₁**] by F. Waldhausen.

There have been successes in higher dimensions: Smale [**S₃**] and Stallings [**S₇**] showed that a compact connected combinatorially triangulated n-manifold M^n is homeomorphic to Bd Δ^{n+1} if $n \geq 5$ and the homotopy groups of M^n are the same as those of Bd Δ^{n+1}. Smale even showed that M^n was combinatorially equivalent to Bd Δ^{n+1}. Related results are given by Wallace [**W₂**] and Zeeman [**Z₁**]. Also, in a preprint Michael Freedman [**F₄**] has treated the 4-dimensional case. While these are very interesting results, we shall not pursue them here since our primary concern here is with 3-manifolds.

XVIII.9. Decompositions. Recall that a subset of R^3 is called *cellular* if it is the intersection of a sequence of PL 3-cells $B_1, B_2, \ldots$, such that $B_{i+1} \subset \text{Int } B_i$ for each i. The approximation theorem (Theorem XIII.A) can be used to show that a set in R^3 is cellular if it is the intersection of a decreasing sequence of crumpled cubes $C_1, C_2, \ldots$, such that $C_{i+1} \subset \text{Int } C_i$ for each i. In studying a closed subset X in R^3, it is frequently convenient to express X as the intersection of PL sets. The approximation theorem is sometimes used in replacing intersections of topological objects with the intersections of polyhedra.

Suppose R^3 is the union of a collection G of mutually disjoint cellular sets such that if an open set U in R^3 contains an element of G, the union of the elements of G in U is open in R^3. Some of the elements of G might be points, and we call each such element degenerate. We regard R^3/G as a topological space whose points are the elements of G and whose open sets are collections of elements of G whose union is open in R^3. We call R^3/G a *cellular decomposition* of R^3. It G has only a finite number of nondegenerate elements, then R^3/G is homeomorphic to R^3.

The decomposition space R^3/G may be shown to have the same homotopy and homology properties (both local and global) as R^3. At one time researchers tried to show that this implied that R^3/G was in fact homeomorphic to R^3. A corresponding 2-dimensional version of this result had been shown by Moore [$\mathbf{M}_{12}$] to hold in R^2. However, in 1957, Bing [$\mathbf{B}_8$] described a cellular decomposition space of R^3 (later called a *dog-bone-space*) such that the space was topologically different from R^3 even though each nondegenerate element of it was a tame arc. Following this, a wide assortment of exotic cellular decompositions of R^3 were constructed. See Armentrout [$\mathbf{A}_9, \mathbf{A}_{10}, \mathbf{A}_{11}$], Armentrout with Bing [**AB**], Bing [$\mathbf{B}_{15}$], Bing with Starbird [**BS**], Eaton [$\mathbf{E}_3$], Fort [$\mathbf{F}_1$], Martin [$\mathbf{M}_1$], Starbird [$\mathbf{S}_8, \mathbf{S}_9$] and Starbird with Woodruff [$\mathbf{SW}_2$]. The topological properties of a cellular decomposition are not determined merely by its algebraic properties.

XVIII.10. Other references. Moise's book [$\mathbf{M}_{12}$] is very relevant to our study. In fact, Moise was one of the pioneers who enabled topologists to use piecewise linear techniques to study geometric 3-manifolds. His work has been an inspiration and his book gives an excellent treatment of many of the topics covered in the present book.

The paper [$\mathbf{BC}_2$] by Burgess and Cannon gives an extensive list of applications of basic material from the present book. The presentation is far more complete than our scant coverage in Chapter XVIII, and the set of references and suggestions for proofs are excellent. Rolfsen [$\mathbf{R}_1$] and Hempel [$\mathbf{H}_6$] contain many excellent treatments.

Our study of 3-manifolds started from an understanding of the geometric topology of the plane. Moore's colloquium volume [$\mathbf{M}_{14}$] contains much valuable information. Hall and Spencer [**HS**] is a good source as is Newman [**N**]. Volume I of Kerèkjàrtò [$\mathbf{K}_1$] was devoted to plane topology. This book is well known for its reference to Bessel-Hagen, but there is another interesting story connected with it. It is said that it was intended that Volume II would treat the topology of 3-manifolds. It is reported that Kerèkjàrtò decided not to write Volume II feeling that at that time in history not enough was known about 3-manifolds to warrant a book. Things are quite different now.

This book considers 3-manifolds from both the piecewise linear and topological points of view but pays scant attention to the differentiable category. If a manifold is given a suitable metric, one can use the resulting structure to get results about derivatives, geodesics, tangents, curvatures, volumes, etc. These results may have interesting applications in the topological category. For example, Bill Thurston uses hyperbolic structures on 3-manifolds to help prove the Smith conjecture—a tame simple closed curve in S^3 is unknotted if it is the fixed point set of a periodic homeomorphism. There is some speculation that similar techniques may lead to a solution of the Poincaré conjecture. Page 501 of the October 1982 *Notices of the American Mathematical Society* contains an article by William Browder and W.-C. Hsiang describing discoveries which led to Thurston being awarded a Field Medal. Pages 293–296 of the August 1979 *Notices of the American Mathematical Society* contain articles by Blaine Lawson and Dennis Sullivan discussing results by Thurston leading to his being awarded the Veblen Prize in Geometry in 1976 and the National Science Foundation's Waterman Award in 1979. Although we regard this direction as outside the scope of this book and do not include references, we do mention that others that have gotten notable results along these and other algebraic lines include Joan Birman, Bill Jaco, John Milnor and Peter Shalen.

We borrowed liberally from material that was primarily aimed at higher dimensions. Perhaps Seifert and Threlfall [ST] is the traditional standard here, but Alexandroff [A_5], Hurewicz and Wallman [HW], Hocking and Young [HY], Hudson [H_{11}], Rourke and Sanderson [RS], Rushing [R_4], Massey [M_3], Wilder [W_7], Whyburn [W_6] and Zeeman [Z_2] are helpful.

SOME STANDARD RESULTS IN TOPOLOGY

In this Appendix we list some standard results. Some of the results follow from methods used in this book. Our reason for putting them in the Appendix was that we wanted to use the results but did not want to pause to give the proofs. Many readers already know the results, and others could prove them after some reflection. Some useful results in topology are included in this Appendix even if they are not specifically used in the book. It is felt that they should be a part of mathematician's cultural heritage and can be useful on related topics. We do not include references nor indications of proof. We do not include definitions that are found in standard sources.

1. Metric spaces. (Some of these results hold in more general spaces but it is not our purpose to pursue that here. For simplicity we may suppose that the underlying space is metric for that is the context in which we use the results. Throughout this section we use M to denote a metric space.) We use the convention that continua are compact and compact sets are closed. We mention that since some references (see [M_{14}]) use different definitions.

1.A. *Baire category theorem.* If M is complete, it is not the union of a countable number of nowhere-dense sets. (A complete metric is one in which Cauchy sequences converge.)

1.B. *Urysohn's lemma.* If A and B are two mutually disjoint closed subsets of M, then there is a map $f: M \to [0, 1]$ such that if $f(A) = 0$, $f(B) = 1$. (In this theorem it is customary to use a normal Hausdorff space rather than a metric space.)

1.C. *Tietze extension theorem.* If Y is a closed subset of M and f is a map of Y into R^1, then f extends to a map of M into R^1. (It is customary to pick X to be normal and Hausdorff rather than metric and the image to be [0, 1] rather than R^1. The image space could be assumed to be R^n.)

1.D. *Arc theorem.* If M is complete and locally connected, each connected open subset of M is arcwise connected and locally arcwise connected. (It is customary to use a Moore space [M_{14}] rather than a metric space for this result. Rather than using a Cauchy sequence to define completeness, one uses the existence of a point in a decreasing sequence of closed sets.)

1.E. *To-the-boundary-theorem.* If U is a proper, open subset of a continuum X in M, then each component of U has a limit point on Bd U.

1.F. *Separation.* Suppose C is a component of a compact set X and A is a closed set in $X - C$. Then X is the union of two mutually disjoint open subsets X_C, X_A containing C and A, respectively.

1.G. *Irreducible connector.* Suppose U_1, U_2 are mutually disjoint open subsets of compact set X such that X is not the union of two mutually disjoint open subsets X_1, X_2 one containing U_1 and the other U_2. Then there is a continuum C in X intersecting both $\overline{U_1}$ and $\overline{U_2}$ such that each of $C - \overline{U_1}$, $C - \overline{U_2}$, and $C - (\overline{U_1} \cup \overline{U_2})$ is connected, but no proper subcontinuum of C intersects both $\overline{U_1}$ and $\overline{U_2}$.

1.H. *Completeness.* If X is the union of a countable collection of closed sets in a complete metric space M, then $M - X$ can be remetrized so as to be complete. We frequently say that each G_δ subset of a complete metric space is topologically complete.

1.I. *Closed separators.* If a set A separates p from q in a connected space M, some closed set in A separates p from q in M. Complete normality rather than metrizable is the crucial property here.

1.J. *Irreducible separator.* If a subset X of a connected and locally connected set M separates p from q in M, then some subset Y of X is irreducible with respect to separating p from q in M. (Y separates p from q in M but no proper subset does.)

1.K. *Collections of separators.* No connected and separable space M contains an uncountable collection of mutually disjoint continua each of which separates M into more than two pieces.

1.L. *Local connectivity.* If a continuum X fails to be locally connected at a point p, it contains a nondegenerate subcontinuum Y such that X is not locally connected at any point of Y.

1.M. *A nonseparating set.* If a continuum X is not locally connected it contains a countable number of points whose union does not separate X.

1.N. *Full normality.* The space M is fully normal.

2. Planar results. We list some useful planar results.

2.A. *Cellularity.* A continuum in R^2 is cellular if it does not separate R^2. (This is the statement of a theorem and is not intended as a definition.)

2.B. *Unicoherence.* R^2 is unicoherent. Similar results hold for Rm and S^m for $m \geqslant 2$. (If R^2 is the union of two closed connected sets A, B, then $A \cap B$ is connected.)

2.C. *Separation with curves.* Suppose in R^2 that C_1, C_2 are different components of a closed set X and C_1 is compact. Then a simple closed curve in $R^2 - X$ separates C_1 from C_2.

2.D. *More separation.* Suppose in R^2 that Y is a finite subset of a closed set X and C_1, C_2 are components of $X - Y$ with C_1 bounded. Then there is a simple closed curve J in $(R^2 - X) \cup Y$ that separates C_1 from C_2.

2.E. *Decreasing sequences of disks.* If in R^2, C is a cellular component of a closed set X, then C is the intersection of a decreasing sequence of disks $D_1^2, D_2^2, D_3^2, \ldots$, such that each lies in the interior of the preceding and each $X \cap \text{Bd } D_i^2 = \varnothing$.

2.F. *Splitting an annulus.* If X is a closed subset of an annulus A such that no component of X intersects both boundary components of A, then there is a simple closed curve in $A - X$ that separates the boundary components of A from each other in A.

2.G. *A minimal disk.* Suppose J_1, J_2 are two simple closed curves in R^2 such that $J_1 \cap J_2$ has more than one point. Then the closure of each bounded component of $R^2 - (J_1 \cup J_2)$ is a disk.

2.H. *Intersections of annuli.* Suppose in R^2 that X is a closed set and C is a compact component of X such that $R^2 - C$ has precisely two components. Then C is the intersection of a decreasing sequence of annuli $A_1^2, A_2^2, \ldots$, in R^2 such that each A_{i+1}^2 lies in the interior of the preceding and no Bd A_i intersects X.

2.I. *Homeomorphisms on annuli.* If J_1, J_2 are two simple closed curves in an annulus A each of which separates the boundary components of A from each other in A, then there is a homeomorphism h of A onto itself such that h is fixed on Bd A and $h(J_1) = J_2$.

2.J. *Expansions.* Suppose Δ^2 is a 2-simplex in R^2, X is a closed set in Δ^2 such that $X \cap \text{Bd } \Delta^2$ is finite, E is an expansion of X in Int Δ^2 with the no-triod property, p and q are limit points of a single component of $X \cap \text{Int } \Delta^2$, and pq is a spanning arc of Δ^2 from p to q in E. Then there is a disk D in $\{p\} \cup \{q\} \cup \text{Int } \Delta^2$ such that Bd $D \subset \text{Bd } E$ and D contains each spanning arc of Δ^2 from p to q in E.

2.K. *Homeomorphic images separate.* If C is a continuum in R^2 and h is a homeomorphism of C into R^2, then $R^2 - C$ and $R^2 - h(C)$ have the same number of components.

2.L. *Nonseparating continua are cellular.* If C is a continuum in R^2 and C^+ is the union of C and the bounded components of $R^2 - C$, then C^+ is cellular in R^2.

3. Results about 2-spheres. In a certain sense the results about 2-spheres could be included with planar results, but we consider them separately. Many of the results depend on the compactness of S^2.

3.A. *Dividing a 2-sphere.* If a subset of S^2 divides S^2 into very small pieces, some continuum in the subset divides S^2 into small pieces. For each 2-sphere S and each $\varepsilon > 0$ there is a $\delta > 0$ such that if X is a subset of S such that the diameter of each component of $S - X$ is less than δ, then X contains a continuum C such that each component of $X - C$ has diameter less than ε.

3.B. *Covering with disks.* If X is a closed subset of a 2-sphere S^2 and the components of X are very small, then there is a finite collection of mutually disjoint small disks on S^2 covering X.

3.C. *Nonhour-glass property.* If X is a very small subset of a 2-sphere S^2, all but at most one component of $S^2 - X$ is small. In fact, given $S^2 \subset R^3$ and $\varepsilon > 0$, there is a $\delta > 0$ such that if h is any δ homeomorphism of S^2 into R^3 and X is any δ-subset of R^3, then at most one component of $h(S^2) - X$ is of diameter more than ε.

3.D. *Kline sphere characterization*. A nondegenerate connected locally connected metric continuum is topologically S^2 if it is separated by each 1-sphere but not by any 0-sphere in it.

3.E. *Separating closed subsets*. If X_i ($i = 1, 2$) is a closed subset of an arc A_i on S^2 such that $X_1 \cap X_2 = \varnothing$, then there is a simple closed curve on S^2 that separates X_1 from X_2 on S^2.

3.F. If G is an uncountable collection of mutually disjoint continua in S^2 each of which separates S^2, then all but at most a countable number of elements of G are irreducible separators.

4. Cylinders in R^3.

In Chapters XI, XII, and XIII we used a right circular cylinder C^3 in R^3 and noted how a 2-sphere K^2 intersected the side of C^3. Recall that L was the straight line through the axis of C^3 and K intersected neither end of C^3.

4.A. *Circling continua*. If X is a circling continua of C^3 in $K^2 \cap$ Bd C^3 and $p \in K^2 - C^3$ then X separates some point of $K^2 \cap L$ from p in K^2.

4.B. *An open subset V of K^2*. If V is a connected open subset of K^2, then Bd V has at most a finite number of components each of which contains a circling continuum of C^3.

4.C. *Covering side-sets with disks*. Suppose X is a closed subset of the side of C^3 such that no components of X intersects either end of C^3.

If W is the union of the components of X that circle C^3, then W is closed. Also there is a sequence of mutually disjoint disks $D_1^2, D_2^2, \ldots$, on the side of C^3 such that the boundaries of the disks miss X, the interiors of the disks cover $X - W$, and $\{D_i^2\}$ is locally finite at each point of Bd $C^3 - W$.

4.D. *Circling continua in disks*. If X is a component of $K^2 \cap$ Bd C^3 that circles C^3, then each disk in K^2 that contains X intersects L.

4.E. *Varying radius of C^3*. If P is plane normal to the axis of C^3 at its center, use C_θ^3 ($0 < \theta \leqslant 1$) to denote the right circular cylinder in R^3 such that C_θ^3 has the same axis as C^3, its ends lie in the end of C^3, but the radius of $P \cap C_\theta^3$ is θ times that of $P \cap C^3$. Then except for a countable number of θ's, each component X of $K^2 \cap C_\theta^3$ has the following properties:

dim $X \leqslant 1$,

if X separates Bd C_θ^3, no proper subset of it does, and

Bd $C_\theta^3 - X$ has at most two components.

5. Decompositions.

While it is possible to describe a decomposition of a space X with the use of a map f of X into another space Y some of the more interesting applications come not from considering a map but rather from considering a collection of mutually disjoint closed sets which satisfy certain upper semicontinuous conditions.

5.A. *Curves on a 2-sphere*. Suppose $\{G\}$ is a collection of mutually disjoint simple closed curves on a 2-sphere S^2. If $S^2 - \cup G$ is finite, it is a 2-point set and each element of G separates the two points from each other on S^2.

5.B. *Finite cellular decomposition*. If $C_1, C_2, \ldots, C_n$ is a finite collection of nondegenerate mutually disjoint cellular subsets of a 2-sphere K^2, then the decomposition space of K^2 having the C's as its set of nondegenerate elements is topologically K^2. In fact, if U is an open set containing $\cup C_i$, there is a map f of K^2 onto itself that is fixed on $K^2 - U$ and has the C's for its set of nondegenerate point inverses.

5.C. *Decomposing a disk*. If G is a monotone upper semicontinuous decomposition of a 2-cell I^2 such that each element of G is a nonseparating proper subcontinuum of I^2 that does not contain Bd I^2, then the decomposition space I^2/G is topologically I^2.

5.D. *Moore decomposition theorem*. If G is a monotone upper semicontinuous decomposition of a 2-sphere K^2, the decomposition space K^2/G is a cactoid. If G is nondegenerate and no element of G separates K^2, the decomposition space is a 2-sphere.

REFERENCES

[A₁] J. W. Alexander, *Theorem on the interior of a simply connected closed surface in three-space*, Bull. Amer. Math. Soc. **28** (1922), 10.

[A₂] _____, *On the subdivision of 3-space by a polyhedron*, Proc. Nat. Acad. Sci. USA. **10** (1924), 6–8.

[A₃] _____, *An example of a simply connected surface bounding a region which is not simply connected*, Proc. Nat. Acad. Sci. USA. **10** (1924), 8–10.

[A₄] _____, *Remarks on a point set constructed by Antoine*, Proc. Nat. Acad. Sci. USA. **10** (1924), 10–12.

[A₅] P. S. Alexandroff, *Combinatorial topology* (English translation, 3 volumes), Graylock Press, Rochester, N.Y., 1956.

[A₆] W. R. Alford, *Some "nice" wild 2-spheres in E^3*, Topology of 3-Manifolds and Related Topics, Prentice Hall, Englewood Cliffs, N. J., 1962, pp. 29–33.

[A₇] _____, *Uncountably many different involutions of S^3*, Proc. Amer. Math. Soc. **17** (1966), 186–196.

[A₈] L. Antoine, *Sur l'homeomorphie de deux figures et de leurs voisinages*, J. Math. Pures Appl. **86** (1921), 221–325.

[A₉] S. Armentrout, *Upper semi-continuous decompositions of E^3 with at most countably many non-degenerate elements*, Ann. of Math. (2) **78** (1963), 605–618.

[A₁₀] _____, *Concerning cellular decompositions of 3-manifolds that yield 3-manifolds*, Trans. Amer. Math. Soc. **133** (1968), 307–332.

[A₁₁] _____, *Cellular decompositions of 3-manifolds that yield 3-manifolds*, Bull. Amer. Math. Soc. **75** (1969), 453–456.

[AB] S. Armentrout and R. H. Bing, *A toroidal decomposition of E^3*, Fund. Math. **60** (1967), 81–87.

[B₁] B. J. Ball, *The sum of two solid horned spheres*, Ann. of Math. (2) **69** (1959), 253–257.

[B₂] R. H. Bing, *The Kline sphere characterization problem*, Bull. Amer. Math. Soc. **52** (1946), 644–653.

[B₃] _____, *Complementary domains of continuous curves*, Fund. Math. **36** (1949), 303–318.

[B₄] _____, *A characterization of 3-space by partitioning*, Trans. Amer. Math. Soc. **70** (1951), 15–27.

[B₅] _____, *A homeomorphism between the 3-sphere and the sum of two horned spheres*, Ann. of Math. (2) **56** (1952), 354–362.

[B₆] _____, *Partitioning continuous curves*, Bull. Amer. Math. Soc. **58** (1952), 536–556.

[B₇] _____, *Locally tame sets are tame*, Ann. of Math. (2) **59** (1954), 145–158.

[B₈] _____, *A decomposition of E^3 into points and tame arcs such that the decomposition space is topologically different from E^3*, Ann. of Math. (2) **65** (1957), 484–500.

[B₉] _____, *Approximating surfaces with polyhedral ones*, Ann. of Math. (2) **65** (1957), 456–483.

[B₁₀] _____, *Necessary and sufficient conditions that a 3-manifold be S^3*, Ann. of Math. (2) **68** (1958), 17–37.

[B₁₁] _____, *An alternative proof that 3-manifolds can be triangulated*, Ann. of Math. (2) **69** (1959), 37–65.

[B₁₂] _____, *Conditions under which a surface in E^3 is tame*, Fund. Math. **47** (1959), 105–139.

[B₁₃] _____, *A wild surface each of whose arcs is tame*, Duke Math. J. **28** (1961), 1–16.

229

[$\mathbf{B_{14}}$] _____, *A surface is tame if its complement is 1-ULC*, Trans. Amer. Math. Soc. **101** (1961), 294–305.

[$\mathbf{B_{15}}$] _____, *Decompositions of E^3*, Topology of 3-Manifolds and Related Topics, Prentice Hall, Englewood Cliffs, N. J., 1962, pp. 5–21.

[$\mathbf{B_{16}}$] _____, *Approximating surfaces from the side*, Ann. of Math. (2) **77** (1963), 145–192.

[$\mathbf{B_{17}}$] _____, *Each disk in E^3 contains a tame arc*, Amer. J. Math. **84** (1962), 583–590.

[$\mathbf{B_{18}}$] _____, *Each disk in E^3 is pierced by a tame arc*, Amer. J. Math. **84** (1962), 591–599.

[$\mathbf{B_{19}}$] _____, *Some aspects of the topology of 3-manifolds related to the Poincaré conjecture*, Lectures in Modern Math., vol. II, Wiley, New York, 1964, pp. 93–128.

[$\mathbf{B_{20}}$] _____, *Inequivalent families of periodic homeomorphisms*, Ann. of Math. (2) **80** (1964), 78–93.

[$\mathbf{B_{21}}$] _____, *Pushing a 2-sphere into its complement*, Michigan Math. J. **11** (1964), 33–45.

[$\mathbf{B_{22}}$] _____, *Improving the side approximation theorem*, Trans. Amer. Math. Soc. **116** (1965), 511–525.

[$\mathbf{B_{23}}$] _____, *Computing the fundamental group of the complements of curves*, Tech. report 2, Washington State Univ., Washington, 1965.

[$\mathbf{B_{24}}$] _____, *Challenging conjectures*, Amer. Math. Monthly (2) **74** (1967), 56–64.

[$\mathbf{B_{25}}$] _____, *Improving the intersections of lines and surfaces*, Michigan Math. J. **14** (1967), 155–159.

[$\mathbf{B_{26}}$] _____, *Models for S^3*, Visiting Scholars Lectures, Texas Tech. University Mathematics Series No. 9, 1970–71.

[$\mathbf{B_{27}}$] _____, *Pulling back feelers*, Symposia Mathematicae **16** (1975), 245–266.

[$\mathbf{B_{28}}$] _____, *Vertical general position*, General Topology, Lecture Notes in Math., vol. 438, Springer-Verlag, Berlin and New York, 1975, pp. 16–41.

[**BM**] R. H. Bing and J. M. Martin, *Cubes with knotted holes*, Trans. Amer. Math. Soc. **155** (1971), 217–231.

[**BS**] R. H. Bing and Michael Starbird, *A decomposition of E^3 with a null sequence of cellular arcs*, Geometric Topology, Academic Press, New York, 1979, pp. 3–21.

[$\mathbf{B_{29}}$] M. Brown, *A proof of the generalized Schoenflies theorem*, Bull. Amer. Math. Soc. **66** (1960), 74–76.

[$\mathbf{B_{30}}$] _____, *Locally flat imbeddings of topological manifolds*, Ann. of Math. (2) **75** (1962), 331–341.

[$\mathbf{B_{31}}$] C. E. Burgess, *Properties of certain types of wild surfaces in E^3*, Amer. J. Math. **86** (1964), 325–338.

[$\mathbf{B_{32}}$] _____, *Characterizations of tame surfaces in E^3*, Trans. Amer. Math. Soc. **114** (1965), 80–97.

[$\mathbf{BC_1}$] C. E. Burgess and J. W. Cannon, *Tame subsets of spheres in E^3*, Proc. Amer. Math. Soc. **22** (1969), 395–401.

[$\mathbf{BC_2}$] _____, *Embeddings of surfaces in E^3*, Rocky Mountain J. Math. (2) **1** (1971), 260–344.

[**BL**] C. E. Burgess and L. D. Loveland, *Sequentially 1-ULC surfaces in E^3*, Proc. Amer. Math. Soc. **19** (1968), 653–659.

[$\mathbf{C_1}$] S. S. Cairns, *An elementary proof of the Jordan-Schoenflies theorem*, Proc. Amer. Math. Soc. **2** (1951), 860–867.

[$\mathbf{C_2}$] J. W. Cannon, *Characterization of taming sets on 2-spheres*, Trans. Amer. Math. Soc. **147** (1970), 289–299.

[$\mathbf{C_3}$] _____, *Sets which can be missed by side approximations to 2-spheres*, Pacific J. Math. **34** (1970), 321–334.

[$\mathbf{C_4}$] _____, *Characterization of tame subsets of 2-spheres in E^3*, Amer. J. Math. **94** (1972), 173–188.

[$\mathbf{C_5}$] _____, *New proofs of Bing's approximation theorems for surfaces*, Pacific J. Math. **46** (1973), 361–379.

[$\mathbf{C_6}$] L. O. Cannon, *Sums of solid horned spheres*, Trans. Amer. Math. Soc. **122** (1966), 203–228.

[$\mathbf{C_7}$] B. G. Casler, *On the sum of two solid Alexander horned spheres*, Trans. Amer. Math. Soc. **116** (1965), 135–150.

[$\mathbf{C_8}$] D. R. J. Chillingworth, *Collapsing three-dimensional polyhedra*, Proc. Cambridge Philos. Soc. **63** (1967), 353–357.

[$\mathbf{C_9}$] R. P. Coelho, *On the group of certain linkages*, Portugaliae Math. **6** (1947), 57–65.

[$\mathbf{C_{10}}$] R. Connelly, *A new proof of Brown's collaring theorem*, Proc. Amer. Math. Soc., **27** (1971), 180–182.

[C_{11}] R. Craggs, *Improving the intersection of polyhedra in 3-manifolds*, Illinois J. Math. **12** (1968), 567–586.

[CF] R. H. Crowell and R. H. Fox, *Introduction to knot theory*, Ginn, Boston, Mass., 1963.

[D_1] R. J. Daverman, *A new proof for the Hosay-Lininger theorem about crumpled cubes*, Proc. Amer. Math. Soc. **23** (1969), 52–54.

[DE_1] R. J. Daverman and W. T. Eaton, *A dense set of sewings of two crumpled cubes yields S^3*, Fund. Math. **65** (1969), 51–60.

[DE_2] _____, *Universal crumpled cubes*, Topologie **11** (1972), 223–235.

[D_2] M. Dehn, *Uber die Topologie des dreidemensionalen baumes*, Math. Ann. **69** (1910), 137–168.

[D_3] P. H. Doyle, *Union of cell pairs in E^3*, Pacific J. Math. **10** (1960), 521–524.

[DH] P. H. Doyle and J. G. Hocking, *Some results on tame discs and spheres in E^3*, Proc. Amer. Math. Soc. **11** (1960), 832–836.

[E_1] W. T. Eaton, *Side approximations in crumpled cubes*, Duke Math. J. **35** (1968), 707–719.

[E_2] _____, *The sum of solid spheres*, Michigan Math. J. **19** (1972), 193–207.

[E_3] _____, *Applications of a mismatch theorem to decomposition spaces*, Fund. Math. (3) **89** (1975), 199–224.

[F_1] M. K. Fort, Jr., *A note concerning a decomposition space defined by Bing*, Ann. of Math. (2) **65** (1957), 501–504.

[F_2] _____, *A wild sphere which can be pierced at each point by a straight line segment*, Proc. Amer. Math. Soc. **14** (1963), 994–995.

[F_3] R. H. Fox, *A quick trip through knot theory*, Topology of 3-Manifolds and Related Topics, Prentice Hall, Englewood Cliffs, N. J., 1962, pp. 120–167.

[FA] R. H. Fox and E. Artin, *Some wild cells and spheres in three-dimensional space*, Ann. of Math. (2) **49** (1948), 979–990.

[F_4] Michael Freedman, *The topology of 4-dimensional manifolds*, 1982, preprint.

[G_1] D. S. Gillman, *Side approximation, missing an arc*, Amer. J. Math. **85** (1963), 459–476.

[G_2] _____, *Note concerning a wild sphere of Bing*, Duke Math. J. **31** (1964), 247–254.

[G_3] W. Graub, *Die semilinearen Abbildungen*, Springer-Verlag, Berlin and New York, 1950.

[G_4] H. C. Griffith, *A characterization of tame surfaces in three space*, Ann. of Math. (2) **69** (1959), 291–308.

[H_1] W. Haken, *On homotopy 3-spheres*, Illinois J. Math. **10** (1966), 159–178.

[HS] D. W. Hall and G. S. Spencer II, *Elementary topology*, Wiley, New York, 1955.

[H_2] A. J. S. Hamilton, *The triangulation of 3-manifolds*, Quart. J. Math. Oxford Ser. (2) **27** (1976), 63–70.

[H_3] O. G. Harrold, Jr., *Some consequences of the approximation theorem of Bing*, Proc. Amer. Math. Soc. **8** (1957), 204–206.

[H_4] _____, *Locally tame curves and surfaces in three-dimensional manifolds*, Bull. Amer. Math. Soc. **63** (1957), 293–305.

[H_5] J. Hempel, *A surface in S^3 is tame if it can be deformed into each complementary domain*, Trans. Amer. Math. Soc. **111** (1964), 273–287.

[H_6] _____, *Free surfaces in S^3*, Trans. Amer. Math. Soc. **141** (1969), 263–270.

[H_7] _____, *3-manifolds*, Ann. of Math. Studies No. 86, Princeton Univ. Press, Princeton, N. J., 1976.

[H_8] D. W. Henderson, *Extensions of Dehn's lemma and the loop theorem*, Trans. Amer. Math. Soc. **120** (1965), 448–469.

[HY] J. G. Hocking and G. S. Young, *Topology*, Addison-Wesley, Reading, Mass., 1961.

[H_9] T. Homma, *On Dehn's lemma for S^3*, Yokohama Math. J. **5** (1957), 223–244.

[H_{10}] N. Hosay, *The sum of a real cube and a crumpled cube is S^3*, Notices Amer. Math. Soc. **11** (1964), 152.

[H_{11}] J. F. P. Hudson, *Piecewise linear topology*, Benjamin, Menlo Park, Calif., 1969.

[HW] H. Hurewicz and H. Wallman, *Dimension theory*, Princeton Math. Series, vol. 4, Princeton Univ. Press, Princeton, N. J., 1941.

[J_1] I. Johannson, *Über singuläre Elementarfläche und das Dehnsche Lemma* I, Math. Ann. **110** (1935), 312–320.

[J_2] _____, *Über singuläre Elementarflächen und das Dehnsche Lemma* II, Math. Ann. **115** (1938), 658–669.

[K₁] B. V. Kerékjártó, *Vorlesungen über Topologie*. I, Flachentopologie, Springer, Berlin, 1923.

[K₂] R. C. Kirby, *Stable homeomorphisms and the annulus conjecture*, Ann. of Math. (2) **89** (1969), 575–582.

[K₃] J. Kister, *Small isotopies in Euclidean spaces and 3-manifolds*, Bull. Amer. Math. Soc. **65** (1959), 371–373.

[K₄] H. Kneser, *Geschlossene Flachen in dreidimensionalen Mannigfaltigkeiten*, Iber. Deutsch. Math.-Verein **38** (1929), 248–260.

[K₅] K. Koseki, *Poincarésche Vermutung in Topologie*, Math. J. Okayama Univ. **8** (1958), 1–106.

[L₁] L. L. Lininger, *Some results on crumpled cubes*, Trans. Amer. Math. Soc. **118** (1965), 534–549.

[L₂] F. M. Lister, *Simplifying intersections of disks in Bing's side approximation theorem*, Pacific J. Math. **22** (1967), 281–295.

[L₃] L. D. Loveland, *Tame surfaces and tame subsets of spheres in E^3*, Trans. Amer. Math. Soc. **123** (1966), 355–368.

[M₁] J. M. Martin, *The sum of two crumpled cubes*, Michigan Math. J. **13** (1966), 147–151.

[M₂] _____, *A rigid sphere*, Fund. Math. **59** (1966), 117–121.

[M₃] W. S. Massey, *Algebraic topology: An introduction*, Harcourt, Brace and World, New York, 1967.

[M₄] D. Mauldin, *The Scottish book*, Birkhauser-Boston, Boston, Mass., 1982.

[M₅] B. Mazur, *On embeddings of spheres*, Bull. Amer. Math. Soc. **65** (1959), 59–65.

[M₆] D. R. McMillan, Jr., *On homologically trivial 3-manifolds*, Trans. Amer. Math. Soc. **98** (1961), 350–367.

[M₇] _____, *Neighborhoods of surfaces in 3-manifolds*, Michigan Math. J. **14** (1967), 161–170.

[M₈] E. E. Moise, *Affine structures in 3-manifolds*. II, *Positional properties, of 2-spheres*, Ann. of Math. (2) **55** (1952), 172–176.

[M₉] _____, *Affine structures in 3-manifolds*. IV, *Piecewise linear approximations of homeomorphisms*, Ann. of Math. (2) **55** (1952), 215–222.

[M₁₀] _____, *Affine structures in 3-manifolds*. V, *The triangulation theorem and Hauptvermutung*, Ann. of Math. (2) **56** (1952), 96–114.

[M₁₁] _____, *Affine structures in 3-manifolds*. VIII, *Invariance of the knot-types; local tame imbeddings*, Ann. of Math. (2) **59** (1954), 159–170.

[M₁₂] _____, *Geometric topology in dimensions 2 and 3*, Springer-Verlag, New York, 1977.

[MS] D. Montgomery and H. Samelson, *A theorem on the fixed points of involutions in S^3*, Canad. J. Math. **7** (1955), 208–220.

[MZ] D. Montgomery and L. Zippin, *Examples of transformation groups*, Proc. Amer. Math. Soc. **5** (1954), 460–465.

[M₁₃] R. L. Moore, *Concerning upper semicontinuous collections of continua*, Trans. Amer. Math. Soc. **27** (1925), 416–428.

[M₁₄] _____, *Foundations of point set theory*, rev. ed., Amer. Math. Soc. Colloq. Publ., vol. 13, Amer. Math. Soc., Providence, R. I., 1962.

[M₁₅] M. Morse, *A reduction of the Schöenflies extension problem*, Bull. Amer. Math. Soc. **66** (1960), 113–115.

[N] M. H. A. Newman, *Elements of the topology of plane sets of points*, Cambridge Univ. Press, Cambridge, Mass., 1954.

[P₁] C. D. Papakyriakopoulos, *On Dehn's lemma and the asphericity of knots*, Ann. of Math. (2) **66** (1957), 1–26.

[P₂] _____, *On solid tori*, Proc. London Math. Soc. (3) **7** (1957), 281–299.

[P₃] _____, *Some problems on 3-dimensional manifolds*, Bull. Amer. Math. Soc. **64** (1958), 317–335.

[P₄] H. Poincaré, *Second complément a l'analysis situs*, Proc. London Math. Soc. **32** (1900), 277–308.

[P₅] _____, *Cinquième complément a l'analysis situs*, Rend. Circ. Mat. Palermo **18** (1904), 45–110.

[R₁] D. Rolfsen, *Knots and links*, Publish or Perish, Berkeley, Calif., 1976.

[R₂] H. Rosen, *Almost locally tame 2-manifolds in a 3-manifold*, Trans. Amer. Math. Soc. **156** (1971), 59–71.

[RS] C. Rourke and B. Sanderson, *Introduction to piecewise-linear topology*, Ergeb. Math. Grenzgeb., vol. 69, Springer-Verlag, New York, 1972.

[**R$_3$**] M. E. Rudin, *An unshellable triangulation of a tetrahedron*, Bull. Amer. Math. Soc. **64** (1958), 90–91.

[**R$_4$**] T. B. Rushing, *Topological embeddings*, Academic Press, New York, 1973.

[**S$_1$**] D. E. Sanderson, *Isotopic deformations of 2-cells and 3-cells*, Proc. Amer. Math. Soc. **8** (1957), 912–922.

[**ST**] H. Seifert and W. Threlfall, *Lehrbuch der Topologie*, Chelsea, 1947.

[**S$_2$**] P. B. Shalen, *A "piecewise linear" proof of the triangulation theorem for 3-manifolds*, dissertation, Harvard University, Cambridge, Mass., 1971.

[**SW$_1$**] A. Shapiro and J. H. C. Whitehead, *A proof and extension of Dehn's lemma*, Bull. Amer. Math. Soc. **64** (1958), 174–178.

[**S$_3$**] S. Smale, *Generalized Poincaré conjecture in dimensions greater than four*, Ann. of Math. (2) **74** (1961), 391–406.

[**S$_4$**] P. A. Smith, *Fixed point theorems for periodic transformations*, Amer. J. Math. **63** (1941), 1–8.

[**S$_5$**] J. R. Stallings, *Uncountably many wild disks*, Ann. of Math. (2) **71** (1960), 185–186.

[**S$_6$**] _____, *On the loop theorem*, Ann. of Math. (2) **72** (1960), 12–19.

[**S$_7$**] _____, *Polyhedral homotopy spheres*, Bull. Amer. Math. Soc. **66** (1960), 485–488.

[**S$_8$**] M. Starbird, *Cell-like, 0-dimensional decompositions of E^3*, Trans. Amer. Math. Soc. **249** (1979), 203–216.

[**S$_9$**] _____, *Null sequence cellular decompositions of E^3*, Fund. Math. **112** (1981), 81–87.

[**SW$_2$**] M. Starbird and E. Woodruff, *Decompositions of E^3 with countably many non-degenerate elements*, Geometric Topology, Academic Press, New York, 1979, pp. 239–252.

[**W$_1$**] F. Waldhausen, *Heegaard-Zerlegungen der 3-sphäre*, Topology **7** (1968), 195–203.

[**W$_2$**] A. H. Wallace, *Modifications and cobounding manifolds*, Canad. J. Math. **12** (1960), 503–528.

[**W$_3$**] J. H. C. Whitehead, *Certain theorems about three-dimensional manifolds*. I, Quart. J. Math. Oxford Ser. **5** (1934), 308–320.

[**W$_4$**] _____, *Three-dimensional manifolds (corrigendum)*, Quart. J. Math. Oxford Ser. **6** (1936), 80.

[**W$_5$**] _____, *On doubled knots*, J. London Math. Soc. **12** (1937), 63–71.

[**W$_6$**] G. T. Whyburn, *Analytic topology*, Amer. Math. Soc. Colloq. Publ., vol. 28, Amer. Math. Soc., Providence, R. I., 1942.

[**W$_7$**] R. L. Wilder, *Topology of manifolds*, Amer. Math. Soc. Colloq. Publ., vol. 32, Amer. Math. Soc., Providence, R. I., 1949.

[**W$_8$**] _____, *A converse of a theorem of R. H. Bing and its generalization*, Fund. Math. **50** (1961/62), 119–122.

[**Z$_1$**] E. C. Zeeman, *The Poincaré conjecture for $n \geqslant 5$*, Topology of 3-Manifolds and Related Topics, Prentice-Hall, Englewood Cliffs, N. J., 1962, pp. 198–204.

[**Z$_2$**] _____, *Seminar on combinatorial topology*, Mimeographed notes, Inst. Hautes Études Sci., Paris, 1963.

INDEX

absolute retract = AR, 298
absolute neighborhood retract = ANR, 298
accessible, 28
Alexander, J. W., 33, 38, 47, 161
Alexander addition theorem, 109
Alexander horned sphere, 34, 38, 41
Alexandroff, P. S., 222
Alford, W. B., 33, 44
almost approximate from, 126, 151
annulus conjecture, 68
Antoine, L., 44, 47
Antoine's necklace, 44
Antoine's wild sphere, 47
approximation theorems, 116, 151, 152, 154
arc theorem, 223
Armentrout, Steve, 221
Artin, E., 33, 47
associativity, 71

Baire category theorem, 223
ball, 2, 33
Ball, B. J., 33, 44
barycentric subdivision, 6
base point, 71
Bessel-Hagen, 221
bicollared, 58
Bing, R. H., 26, 33, 37, 43, 53, 95, 102, 116, 121, 132, 151, 212, 214, 215, 219, 220, 221
Birman, Joan, 222
Borsuk's theorem, 101
bounding cycle, 105
bounds, 105
branch point, 75, 184
brick partitioning, 26
Browder, William, 222
Brown, Morton, 58, 65
Burgess, C. E., 217, 222

Cairn's Stewart S., 19
Cannon, J. W., 217, 221

Cannon, L. O., 33, 44, 217
canonical collared Schoenflies theorem, 62, 73
canonical normal disk, 206
canonical n-sphere, 33, 197
Carathéodory, C., 19
cartesian product neighborhood, 58
Casler, B. G., 33, 44
cell, 2
cellular, 49, 224
cellular decomposition, 221
cellular partitioning, 219
cellular shelling, 170
cellular subdivision, 80, 170
center push, 143
chain, 105
Chillingsworth, D. R. J., 174
circles, 45, 133, 137, 226
Coelho, R. P., 45
collapse, 173
collar, 58
compact support, 13
compatible, 17
complete, 224
complex, 1
coning, 6
conjugacy class, 204
Connelly, Robert, 65
contains most, 151
converges, 140
covering, 70, 225, 226
Craggs, R., 132
crosses, 116
Crowell, R. H., 95
Crumpled cube, 34
curvilinear triangulation, 2
cycle, 105

Daverman, R. J., 33, 121

235

ABCDEFGHIJ–CM–89876543